MASTERING THE CPAT: A COMPREHENSIVE GU

MASTERING THE CPAT: A COMPREHENSIVE GUIDE

Al Wasser and Donna Kimble

THOMSON
DELMAR LEARNING™

Australia Canada Mexico Singapore Spain United Kingdom United States

Mastering the CPAT: A Comprehensive Guide

Al Wasser and Donna Kimble

Vice President, Technology and Trades ABU:
David Garza

Director of Learning Solutions:
Sandy Clark

Acquisitions Editor:
Alison Pase

Product Manager:
Jennifer A. Starr

Marketing Director:
Deborah S. Yarnell

Channel Manager:
Erin Coffin

Marketing Coordinator:
Mark Pierro

Director of Production:
Patty Stephan

Senior Production Manager
Larry Main

Content Project Manager:
Mary Beth Vought

Technology Project Manager:
Kevin Smith

Technology Project Specialist:
Linda Verde

Editorial Assistant:
Maria Di Cerbo Conto

Printed in the United States of America
1 2 3 4 5 XX 08 07 06

For more information contact
Thomson Delmar Learning
Executive Woods
5 Maxwell Drive, PO Box 8007,
Clifton Park, NY 12065-8007
Or find us on the World Wide Web at
www.delmarlearning.com

Title page photo: Courtesy, Simulaids, Inc.

Library of Congress Cataloging-in-Publication Data:
Wasser, Al.
Mastering the CPAT : a comprehensive guide / Al Wasser and Donna Kimble.
p. cm.
Includes index.
ISBN 1-4180-1229-7
1. Fire fighters—Certification. 2. Fire extinction—Examinations—Study guides. I. Kimble, Donna. II. Title. III. Title: Mastering the Candidates physical ability test.
TH9157.W37 2006
628.9'25076—dc22

Card Number:
2006019973

ISBN-13: 978-1-4180-1229-8

ISBN-10: 1-4180-1229-7

DEDICATION

Donna Kimble
I would like to thank Kris for her encouragement and belief in me, and my mom for her support of all that I do. I would also like to thank Al for inviting me to be a part of this project.

Al Wasser
To Paul Grant, Director of the Fire Science Program at Red Rocks Community College, for providing the inspiration and encouragement for this project.

To Jennifer Scott, Lieutenant, Cunningham Fire Department, for thinking "out-of-the-box" to help conceive and develop our CPAT training class.

To Renie Del Ponte, Dean of Instruction, Red Rocks Community College, for providing the support and guidance in this innovative environment.

ABOUT THE AUTHORS

Donna Kimble is a nineteen-year veteran of the City of Westminster Fire Department in Colorado. She is currently a Fire Lieutenant and active in their department fitness program. Donna is an ACE (American Council on Exercise) certified Personal Trainer and a certified Peer Fitness Trainer. She also holds a Bachelor of Science in Recreation Administration and is a certified Paramedic. Donna has a passion and concern for firefighter health and fitness and hopes to be a positive and contributing influence in this area.

Al Wasser is the Fitness Coordinator at Red Rocks Community College in Lakewood, Colorado. He has a Master's degree from Colorado State University and is an American Council on Exercise (ACE) Gold certified personal trainer. He is also a Certified Strength and Conditioning Specialist (C.S.C.S.) from the National Strength and Conditioning Association (NSCA). Al helped develop and instruct the CPAT training program at Red Rocks Community College.

PREFACE

INTENT OF THIS BOOK

This manual is for fire science students who are enrolled in a Candidates Physical Ability Test (CPAT) preparation course. This manual is also for individuals who are required to pass the CPAT as a condition of employment as a firefighter.

The CPAT was developed by the joint effort of the International Association of Firefighters (IAFF) and the International Association of Fire Chiefs (IAFC).

It is part of the Joint Labor and Management Wellness-Fitness Initiative (WFI), a program that seeks to improve the health and fitness capabilities of firefighters.

There has been an emerging trend to require the CPAT as a condition of employment for firefighters. This has led to a growing demand for fire science academies nationwide to offer a CPAT preparation course. However, the lack of a comprehensive preparation guide has forced colleges to develop their own course without a point of reference. Individuals who wish to pass the CPAT have had to rely upon the advice of others who have taken the test. This manual was written as a response to these needs.

Approach of This Book

This book was developed following a comprehensive approach designed to "leave no stone unturned" in an effort to prepare candidates for the test.

Modeled after the training programs of Olympic competitors and triathletes, the text includes training in the knowledge, skills, and abilities (KSAs) needed to pass the CPAT.

KSAs crucial to passing the CPAT are:

Knowledge of the CPAT events and rules
Training in each CPAT event
Mental focus and preparation for each event
Cardiovascular conditioning
Total body muscular strength
Core body strength
Balance training
Proper nutritional habits

Organization of This Book

The text is designed in a sequential format with each chapter building upon the knowledge and skills developed in the previous chapter(s).

- **Chapter 1** covers how the test was developed and the role it plays in the Wellness and Fitness Initiative (WFI). The rationale and purpose of each event is described, with the rules, procedures, equipment used, and reasons for failures listed.
- **Chapter 2** covers the four success principles of effective program design. Goal setting, learning, practice, and feedback are defined and the function that each plays in effective skill development is shown.
- **Chapter 3** is divided into four learning modules: fitness testing, nutrition, training principles, and mental focusing abilities. These modules will provide the student with starting points and direction for developing their learning and practice goals.
- **Chapter 4** is divided into eight learning modules that cover the approach and strategy involved in each CPAT event. Ideal and alternate training exercises are described, and the additional skills of core body and balance training are presented.
- **Chapter 5** presents the concept of periodization, which allows you to schedule your training in a progressive manner so that you will "peak" at the time of the CPAT test. The treatment of injuries and over-training conditions are also outlined.

Features of This Book

This book is designed to prepare you for optimal performance on the CPAT events. As such, many helpful tools are included to ensure the most efficient planning and practice based on your individual needs:

- **Performance Points** that are interspaced throughout the text highlight expert advice on training and provide tips for success.

- **A Performance Planning Chart** accompanies each learning and event module to help you set learning and practice goals and detail how feedback will be received for each event. In turn, it allows you to evaluate your progress and adjust or set new goals if necessary.
- **A "Key Stats" Section** that also accompanies each learning and event module includes actual average times of successful candidates at each event, average times between events, cardiovascular demand, muscles used, and skills needed. Actual average times of successful candidates and average times between events were derived from a sample of 82 candidates on five different testing dates.
- **A Focus on the Periodization Concept** explains how to plan your training through a steady progression of challenging exercises. This allows you to be in peak condition for the test.
- **A Chapter on Nutrition** emphasizes the role that your diet plays in your performance in daily activities, exercise, and the CPAT. Recommendations for various foods vital to a successful performance in practice and on the Test help you prepare accordingly.
- **Case Studies** presented at the end of Chapters 2 through 5 show how chapter concepts are actually applied by candidates.
- **Appendices** cover training programs of varying length, including 16 and 12 week semester plans and eight and four week programs for the individual.

Supplements to This Book

An e.resource CD is available to instructors, and includes the following:

- Answers to Review Questions
- PowerPoint presentations
- Test Bank
- Image Library
- Electronic versions of the Performance Planning Charts

ACKNOWLEDGMENTS

We wish to sincerely thank the Thomson Delmar Learning project team who worked to transform the manuscript into a reality—Alison Pase, Acquisitions Editor; Jennifer Starr, Product Manager; and Mary Beth Vought, Content Project Manager.

Special thanks also to the photographer: Amy Glickson www.pixlstudio.com, whose time and dedication to the photos brought the manual to a new level of quality.

And certainly, we extend our gratitude to the content experts who carefully reviewed the manual and provided insightful recommendations:

Ben Andrews
Assistant Chief
Clallam County Fire District #3
Sequim, WA

Dennis Childress
Fire Captain
Orange County Fire Authority
Irvine, CA

Craig Hanna
Sioux Falls Fire Rescue
Siux Falls, SD

Robert Klinoff
Chief Deputy
Kern County Fire Department
Bakersfield, CA

Steve Malley
Fire Academy Director
Weatherford College Regional Fire Academy
Weatherford, TX

W. G. Shelton, Jr.
Branch Chief
Virginia Department of Fire Programs
Glenn Allen, VA

Andrea Walter
Firefighter/Technician
Metropolitan Washington Airports Authority Fire Department
Washington DC

CONTENTS

THE CPAT: HISTORY AND DEVELOPMENT

LEARNING OBJECTIVES

Upon completion of this chapter the student will be able to:

- List the two organizations that came together to write the Fire Service Joint Labor Management Wellness-Fitness Initiative (WFI).
- List and explain the five major topics of the WFI.
- Explain why the Candidate Physical Ability Test (CPAT) was created and implemented.
- List the eight events involved in the CPAT.
- Explain how the CPAT is validated and why this is important.
- Know the equipment used, skills tested, warnings and failures issued, and proper execution of each event.

THE JOINT LABOR MANAGEMENT WELLNESS-FITNESS INITIATIVE

The Fire Service has a vested interest and concern for the wellness and fitness of its members. "Firefighting is one of the most physically and mentally challenging jobs in North America,"[1] with a high risk for death or injury. Healthy and fit firefighters are able to perform their jobs better, able to handle stress better, and create cost savings by taking fewer sick days and filing fewer workers compensation claims. The number one killer of firefighters is heart attack. Sprains and strains are the most common injury. The most injured parts of the body are the back and knees, often career-ending injuries. Even though it would appear that all firefighters should be in excellent physical condition, this is not always the case. In 1996, the **International Association of Fire Chiefs (IAFC)** and the **International Association of Fire Fighters (IAFF)** came together to write and implement the Joint Labor Management Wellness-Fitness Initiative (WFI). The IAFC represents over 12,000 fire chiefs and fire department chief officers, and the IAFF represents over 274,000 professional firefighters and emergency personnel. The goal of the WFI is to promote wellness and fitness as a job qualification and a lifestyle, to improve the quality of life, and to prolong a quality career and life for all firefighters **(Figure 1-1)**. To accomplish these goals the WFI focuses on five important topics: medical evaluation, fitness testing and exercise, rehabilitation, behavioral health promotion, and data collection.

Figure 1-1 A well-rounded program of exercise, good nutrition, stretching, counseling, and record-keeping will lead to success in the CPAT.

Medical Evaluation

This component of the WFI requires that all uniformed personnel have a yearly physical examination by a physician. The physician will determine whether the individual is physically capable of performing his/her duties. The yearly physical will also detect any changes in health or any patterns of disease, and monitor ongoing problems. All fire departments have medical standards for new hires and **incumbent firefighters** (those currently in the position). The **National Fire Protection Association (NFPA)** developed Standard 1582, *Standard on Comprehensive Occupational Medical Program for Fire Departments*. Departments and jurisdictions that have **adopted** Standard 1582 for their own use require that candidates meet all the medical requirements set forth in this document. If a candidate does not meet the medical requirements, it does not matter how well they do in the rest of the recruiting process. Before beginning any new or different exercise program, an individual should have a physical assessment by a physician. Proactive and preventive health measures are beneficial to detecting and conquering life-threatening diseases.

Fitness Testing and Exercise

All fire department members should have an annual fitness assessment. The **Candidate Physical Ability Test (CPAT)** is a physical test created to evaluate a new candidate's readiness to meet the high physical demands of a firefighting career. Although the CPAT is not designed for incumbent firefighters, it can be used as a measuring stick to measure ongoing fitness levels.

Rehabilitation

After an injury or illness it is imperative that the firefighter is totally rehabilitated and ready to return to the demands of the job. If a physician releases the firefighter back to duty, he/she has to be capable of performing the task and duties required.

Behavioral Health Promotion

A wellness program must include the mental health of its participants. Firefighting is a stressful job and behavioral health has a direct effect on job performance and physical health. Fitness programs have proven to be beneficial in managing stress and maintaining mental health. Behavioral health issues need to be addressed or they can lead to other significant problems, such as drug/alcohol abuse, family problems, or depression. Many departments offer **Employee Assistance Programs** (formal counseling services that cover a range of issues) and **Critical Incident Stress Debriefing Programs** (informal counseling services to assist after traumatic incidents).

Data Collection

The WFI includes a component that addresses the collection of information to determine if the WFI is meeting its goals. Information is to be collected in a uniform and consistent manner and then uploaded into the International Association of Firefighters Data Base. All personal health data collected is kept confidential.

CREATING THE CANDIDATE PHYSICAL ABILITY TEST

After the WFI was established the IAFC and the IAFF worked with a task force of 10 fire departments from the United States and Canada, and a technical committee consisting of labor officials, firefighters, line officers, training officers, attorneys, physicians, and two medical specialists, a **kinesiologist** and an exercise physiologist. The task force included women and racial minorities as well. Together they all developed the Candidate Physical Ability Test. The CPAT is a minimum standard of necessary physical attributes that are needed to perform various firefighting tasks.

This physical assessment was to be implemented in the 10 different fire jurisdictions with the goal of measuring the physical capabilities of new firefighter candidates.

The task force included these 10 fire departments in the United States and Canada:

- Austin, Texas Fire Department
- Calgary, Alberta Fire Department
- Charlotte, North Carolina Fire Department
- Fairfax County, Virginia Fire and Rescue Department
- Indianapolis, Indiana Fire Department
- Los Angeles County, California Fire Department
- Miami-Dade, Florida Fire Rescue Department
- City of New York, New York Fire Department
- Phoenix, Arizona Fire Department
- Seattle, Washington Fire Department

The goal of the IAFF and IAFC in creating the CPAT was to have fair and **valid** testing standards, find quality individuals for the fire service, and allow for diversity in the fire service. The CPAT was developed to measure a candidate's ability to perform common tasks involved in firefighting and to do these tasks in a realistic amount of time. The task force considered many factors in creating the CPAT, including what functions are most critical to the firefighter and his/her job, the equipment a firefighter must use, the weight of various equipment, the firefighter job description, and a detailed job analysis. The conclusion of the task force was to include eight different physical tasks to be performed consecutively.

1. Stair Climb
2. Hose Drag
3. Equipment Carry
4. Ladder Raise and Extension
5. Forcible Entry
6. Search
7. Rescue
8. Ceiling Breach and Pull

A pass/fail time of 10 minutes and 20 seconds was established after reviewing videos of firefighters performing tasks involved in the test. To legally defend the CPAT and make it a legitimate assessment tool it had to be legally validated. The CPAT does meet the validity criteria established by the United States Department of Labor (DOL), Department of Justice (DOJ), Equal Employment Opportunity Commission (EEOC), and the Canadian Human Rights Commission (CHRC). The test is valid because it predicts a candidate's ability to perform on the fire ground and on tasks which are similar to those performed on the job. The task force also wanted to assure that the CPAT could be taken or given in any jurisdiction and used as a valid assessment and tool for hiring quality individuals. For the CPAT to remain valid it must not be changed or added to by individual fire departments, and all aspects of the CPAT manual must be adopted and followed. When a department decides to implement the CPAT they must follow all of the six major components listed below.

Component 1—Recruiting and Mentoring

Fire departments should reflect the diversity of the community they serve. The CPAT promotes diversity and only bases its assessment on the physical capability of the candidate to pass in the allotted time, without giving them a ranking. The fire service should actively pursue recruiting men and women of all ethnic and cultural backgrounds to become members of their departments.

Component 2—CPAT Preparatory Guide

To make the CPAT a fair process for everyone, candidates should be given a copy of the **CPAT Preparation Guide** at least 8 weeks prior to testing. The Preparation Guide describes the eight different stations, the physical demands made upon the candidate, and basic exercise and hydration techniques.

To allow everyone an equal opportunity to become a firefighter, the U.S. EEOC recommends that all participants be provided the same information and opportunities prior to any testing. The Preparation Guide explains the procedures and skills necessary and gives all candidates the benefits of being prepared for the assessment, without regard to their physical capabilities or readiness for testing.

Component 3—CPAT Administration and Orientation

The ***CPAT Manual*** outlines strict procedures to follow when administering the CPAT. To ensure that all candidates are treated equal and fairly, all procedures must be followed, every time, without deviation. The areas covered in the *CPAT Manual* include:

- Logistical and environmental factors
- Venue and test props and equipment
- Scheduling, staging, and support
- Pre-test orientation
- Re-testing

The CPAT guidelines are so specific that they even state what the outside temperature must be to conduct the assessment (between 45°F/7°C and 95°F/35°C). The orientation is a good opportunity for candidates to have all their questions answered and assure themselves that they understand the testing procedures. Candidates will have the opportunity to see and test the equipment used in the assessment, view the test events, and be informed about re-testing procedures. The CPAT allows for fire departments to have a consistent and validated physical standards assessment for hiring new recruits. Even though the CPAT is designed for new hire candidates, it is a good minimum standard of physical fitness and ability in general, which all in-service fire department members should strive to maintain.

Component 4—The Candidate Physical Ability Test

The CPAT is a timed assessment consisting of eight consecutive events with a pass/fail time of 10 minute and 20 seconds. To simulate the wearing of firefighting gear and self-contained-breathing-apparatus (SCBA), the candidate will wear a 50-lb (22.68 kg) weight vest throughout the assessment. To simulate the weight of a **high-rise pack** (a bundle of fire hose) being carried upstairs, the candidate will add an additional 25 lbs (11.34 kg) to the weight vest (for a total of 75 lbs/34.02 kg) for the Stair Climb event. A detailed description of the CPAT events will be discussed later. The *CPAT Manual* discusses the exact procedure for the assessment to be administered, which allows for consistency and fairness for all individuals. All departments administering the CPAT must follow the same standard procedures, without deviation.

Component 5—The CPAT Validation Process

The CPAT has met the validation criteria for the U.S. DOL, DOJ, EEOC, and the CHRC. In validating the CPAT, nearly one thousand surveys regarding firefighting functions and qualifications were collected from the participating departments, with diversity in men and women, age, ethnic background, and demographics. To prevent any legal problems, the *CPAT Manual* must be implemented as it is written and not added to or have any part left out or changed in any way.

Component 6—Legal Issues

In today's climate legal challenges must be a consideration and something to be approached proactively to avoid potential problems. The task force was proactive and made the CPAT consistent with two Federal statutes: Title VII of the Civil Rights Act of 1964, and Title I of the Americans with Disabilities Act of 1990. The best way for legal issues to be avoided is for the assessment administrator to read and follow the *CPAT Manual*. Any deviation from the manual procedures opens the administrator up to questioning and controversy.

RULES AND PROGRESSION OF EACH EVENT

The CPAT is a series of eight events performed in a predetermined sequence, and must be completed in a time no more than 10 minutes and 20 seconds. To make the assessment fair all candidates must wear the same type of clothes and equipment throughout the events. Each person must wear "long pants, a hard hat with a chin strap, work gloves, and shoes with no open heel or toes."[2] For safety the candidate is not allowed to wear jewelry or watches, and it is a good idea to pull back long hair so it does not get in the way. The course will be laid out so that there is an 85-ft (25.91 m) walk between each event—running is not permitted—which allows for the candidate to recover for approximately 20 seconds between events. (See Appendix D for a sample layout of the course.) Each candidate will wear a 50-lb (22.68 kg) weight vest throughout the assessment, with exception of the Stair Climb. During the Stair Climb an additional 25 lbs (11.34 kg) is added to the vest for a total weight of 75 lbs (34.02 kg). The weight vest is used to simulate the added weight a firefighter wears during a fire incident, which includes the turnout gear, SCBA, high-rise pack, or loose tools they may carry. The candidate should be mentally and physically ready to go the day of the assessment.

The remainder of this chapter will discuss each event, describing the equipment needed, which skills are tested, performance guidelines, and what actions will merit a warning or a failure.

EVENT 1—Stair Climb

Equipment Used

- Step Mill (brand of stair climbing machine approved for CPAT use). One handrail is removed and the other handrail is left in place to assist in mounting and dismounting
- Total 75-lb (34.02 kg) weight vest **(Figure 1-2)**

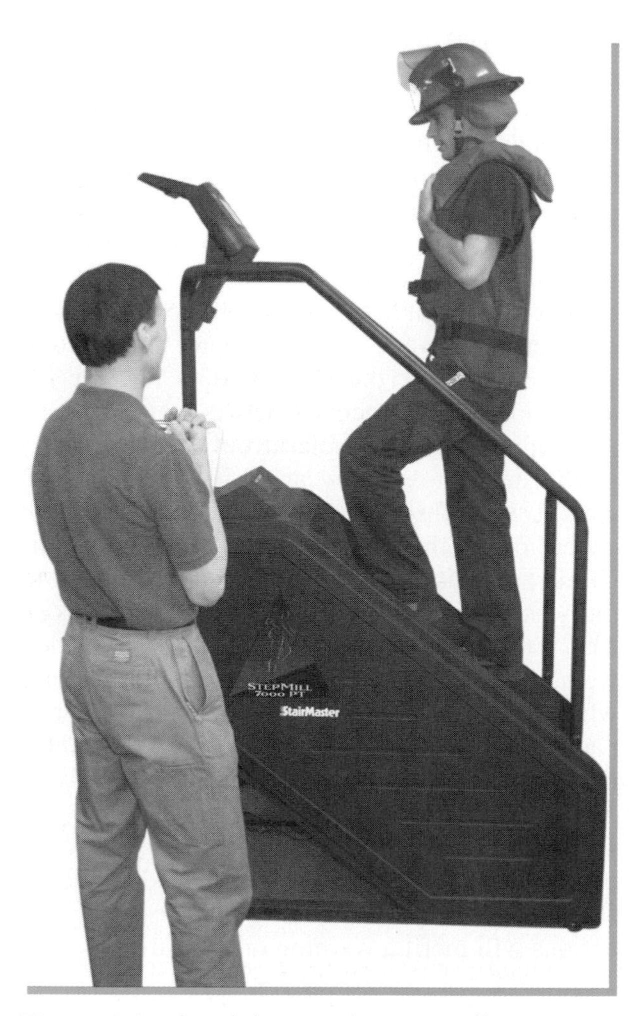

Figure 1-2 Candidate on the step mill.

Skills Being Assessed

This event will simulate the candidate's ability to climb stairs while carrying the weight of a high-rise pack or fire equipment.

Performing the Event

The candidate is given 20 seconds on the Step Mill to warm up. During these 20 seconds the Step Mill is set at 50 steps per minute; the candidate is allowed to dismount if necessary, and he/she is allowed to grasp the handrail to establish balance and establish a comfortable pace. If the candidate accidentally falls off the Step Mill during the 20-second warm-up period, he/she can start over which will restart the 20-second warm-up. The candidate is only allowed to restart twice. Once the warm-up period is completed the assessment time immediately begins, without pause or break. During the actual test the Step Mill will be moving at a rate of 60 steps per minute and will continue for 3 consecutive minutes until the event is completed.

Warnings or Failures

- If the candidate falls or steps off the Step Mill three times during the warm-up period, that is a failure.
- If the candidate falls off the Step Mill during the 3-minute assessment period, the candidate fails.
- If the candidate grasps the handrail or any piece of the equipment during the 3-minute assessment, that is a failure. The rail or wall can be touched momentarily for balance but cannot be grasped onto; if the equipment is touched for balance the candidate will be given a warning, the third warning will be considered a failure and the candidate will not be allowed to continue.

EVENT 2—Hose Drag

Equipment Used

- 200 ft (60 m) of synthetic double-jacketed $1^3/_4$-in. (44-mm) fire hose **(Figures 1-3, 1-4)**
- Nozzle, approximately 6 lbs (3 kg)
- 50-lb (22.68 kg) weight vest

Figure 1-3 Candidate moving a hose line.

Figure 1-4 Candidate pulling a hose line.

Skills Being Assessed

The candidate will simulate dragging an **uncharged** (no water) hose line to a fire, and then pulling the hose hand-over-hand around obstacles while remaining stationary.

Performing the Event

The hose is marked 8 ft (2.44 m) behind the nozzle, the candidate will pull the nozzle and hose over their shoulder and across the front of the body, but not exceed the 8-ft mark. The candidate can walk or run, dragging the hose for 75 ft (22.86 m) until they come to a large drum, take a 90-degree turn around the drum, and continue to drag the hose another 25 ft (7.62 m). Then he/she stops in a marked box, 5 ft (1.52 m) by 7 ft (2.13 m), and drops down on at least one knee and pulls 50 ft (15.24 m) of additional hose toward him/herself. The hose will be marked at the 50-ft (15.24 m) point that the candidate must reach; once that mark is reached the event is over.

Warnings and Failures

- The candidate must stay inside the marked path; if he/she goes outside the path, that is a failure.
- If the candidate does not go around the drum, that is a failure.
- If the candidate has exceeded the 10:20 time limit, that is a failure.
- The candidate must keep at least one knee in contact with the ground or he/she will get one warning; if a second warning is given, that is a failure.
- While pulling the hose the candidate must stay within the marked box at all times or they get a warning; two warnings is a failure.

EVENT 3—Equipment Carry

Equipment Used

- A rescue circular saw weighing approximately 32 lbs (14.5 kg) (**Figure 1-5**)
- A chain saw weighing approximately 28 lbs (12.7 kg)
- Tool cabinet
- 50-lb (22.68 kg) weight vest

Figure 1-5 Candidate performing the Equipment Carry event.

Skills Being Assessed

The candidate is assessed on their ability to remove equipment from the fire **apparatus** (truck), carry the equipment to the fire scene, and then return the equipment to the fire apparatus.

Performing the Event

The candidate will remove the two saws out of a tool cabinet and place them on the ground one at a time. Then he/she will pick a saw up in each hand and carry them 75 ft (22.86 m), go around a large drum 180 degrees, and carry the saws back to the start line. Then the candidate will put both saws on the ground and one at a time replace them into the cabinet. The total distance the candidate will carry the saws is 150 ft (45.72 m). It is acceptable to place the saws on the ground and re-grip them during the assessment.

Warnings and Failures

- If the candidate drops one or both of the saws while carrying them, that is a failure.
- The candidate must walk while doing this event; if they run, they will be given a warning; the second warning is a failure.
- If the candidate exceeds the 10:20 time limit, that is a failure.

EVENT 4—Ladder Raise and Extension

Equipment Used

- Two 24-ft (7.32 m) aluminum extension ladders **(Figures 1-6, 1-7)**
- Attaching brackets for the ladder extension
- 50-lb (22.68 kg) weight vest

Skills Being Assessed

The candidate will simulate raising the ladder to a structure at a fire scene and then extending the ladder to a roof or window.

Performing the Event

The candidate will start at the top rung (rung farthest from the structure) of the 24-ft (7.32 m) extension ladder. The opposite end of the ladder is hinged at the ground in front of the structure. The candidate will grab the top rung and lift the ladder up and over his/her head and walk the ladder, by grabbing each rung, up toward the structure until the ladder is vertical and stationary against the wall. The rails (sides) of the ladder cannot be used while raising the ladder. The candidate then proceeds to another 24-ft (7.32 m) extension ladder, which is secured into position, and stands in a marked box area. The candidate extends the ladder with the rope lanyard until the ladder stops at the top and is fully extended. The candidate then lowers the smaller, movable part of the ladder, the **fly section**, in a controlled hand-over-hand manner back to the start position to complete the event.

Warnings and Failures

- The candidate must use every rung while raising the ladder to the structure; if a rung is missed, the candidate is given a warning; the second warning is a failure.
- If, while extending the fly section of the ladder, the candidate drops the ladder or loses their grip, that is a failure.
- If the candidate cannot fully extend the ladder, he/she fails.
- The candidate must keep his/her feet in the marked box at all times or a warning is given; two warnings is a failure.
- On lowering the fly section of the ladder the candidate must demonstrate complete control and use a hand-over-hand method; if control is lost and the fly section falls, that is a failure.
- If the candidate exceeds the 10:20 time limit, that is a failure.

Figure 1-6 Candidate raising the ladder.

Figure 1-7 Candidate extending the ladder.

EVENT 5—Forcible Entry

Equipment Used

- Forcible entry machine (**Figure 1-8**)
- 10-lb (4.54 kg) sledgehammer
- 50-lb (22.68 kg) weight vest

Figure 1-8 Candidate performing the Forcible Entry event.

Skills Being Assessed

This event will assess the candidate's ability to use force and power to open a door or break through a wall. Forcing doors is a common task that firefighters must be able to perform.

Performing the Event

The candidate will stand outside of a toe-box area in front of the forcible entry machine. Using a 10-lb sledgehammer he/she strikes a target on the machine until a buzzer signal is activated. Once the signal is activated the event is completed.

Warnings and Failures

- If the candidate steps inside the toe-box area, a warning is given; the second warning is a failure.
- If the candidate releases the sledgehammer with both hands, they have failed.
- If the candidate exceeds the 10:20 time limit, that is a failure.

EVENT 6—Search

Equipment Used

- Search maze (**Figure 1-9**)
- 50-lb (22.68 kg) weight vest

Figure 1-9 Candidate performing the Search event.

Skills Being Assessed

This event assesses the candidate's ability to search for a victim in a challenging environment under limited visibility. This search will simulate performing a rescue on the fire scene in smoky conditions.

Performing the Event

This event is done on the candidate's hands and knees, crawling through a tunnel maze. The tunnel maze is approximately 3 ft (91.44 cm) high, 4 ft (121.92 cm) wide, and 64 ft (19.51 m) in length, and includes two 90-degree turns. While crawling through the maze the candidate will have to navigate over and around obstacles, and in two locations the tunnel will narrow. Once the candidate is out of the maze the event is over. If the candidate needs to evacuate the tunnel for any reason, he/she can rap on the tunnel and will be assisted out.

Warnings and Failures

- If the candidate needs assistance or evacuates out of the tunnel before completing the event, that is a failure.
- If the candidate exceeds the 10:20 time limit, that is a failure.

EVENT 7—Rescue

Equipment Used

- One 165-lb (74.84 kg) mannequin (unclothed) (Figure 1-10)
- Mannequin harness
- 50-lb weight vest (for candidate)

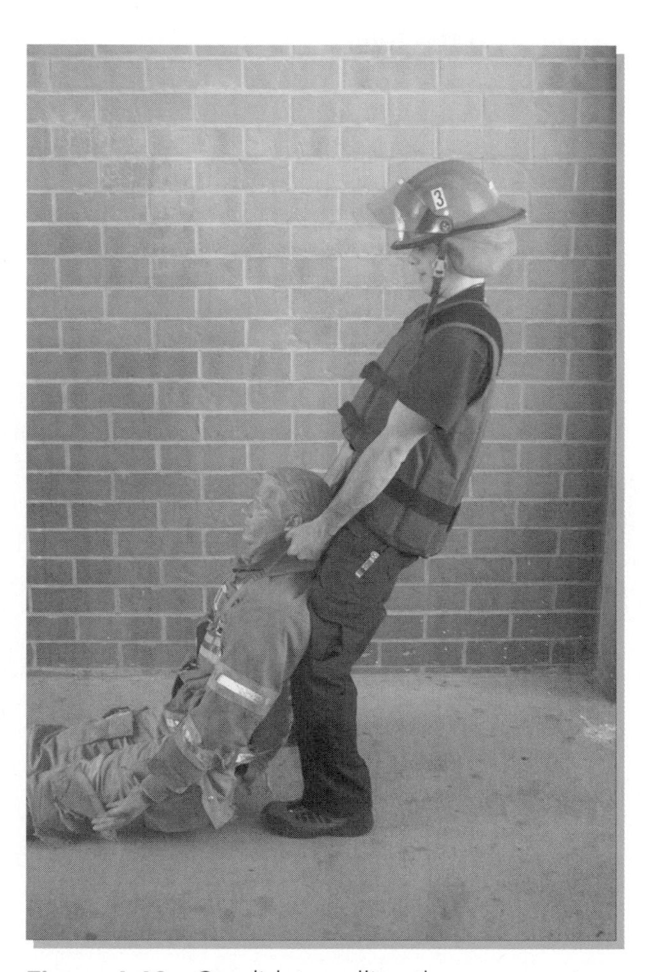

Figure 1-10 Candidate pulling the mannequin.

Skills Being Assessed

This event will simulate the candidate's ability to remove a victim or an injured partner from a hazardous situation.

Performing the Event

Using one or both hands, the candidate grabs the attached harness and drags the mannequin 35 ft (10.67 m) to a drum, turns 180 degrees around the drum, and back the 35 ft (10.67 m) to the finish line. The entire mannequin must be pulled across the line. The candidate cannot touch the drum but the mannequin can touch the drum. The candidate is allowed to stop, re-grip the harness, and continue.

Warnings and Failures

- If the candidate touches the drum, he/she is given a warning; two warnings is a failure.
- If the candidate exceeds the 10:20 time limit, that is a failure.

EVENT 8—Ceiling Breach and Pull

Equipment Used

- Ceiling breach and pull device (**Figure 1-11**)
- 6-ft (1.83 m) pike pole
- 50-lb (22.68 kg) weight vest

Figure 1-11 Candidate breaching the ceiling.

Skills Being Assessed

This event assesses the candidate's ability to use the hooked **pike pole** to break through a ceiling and pull it down. This simulates what a firefighter must do to pull a ceiling to check for fire extension.

Performing the Event

First the candidate removes the pike pole from its bracket and stands within the equipment frame boundary. The tip of the pike pole is placed on a painted area of a hinged door in the ceiling. The candidate pushes up the hinged door until it contacts the top of the apparatus three times. The candidate then hooks the pike pole to the ceiling device and pulls down five times. This completes one set of three pushes and five pulls. The candidate will complete four sets. It is acceptable to re-grip the pike pole or lose the grasp and re-grip as long as the pike pole does not hit the ground. If a repetition is not successfully completed the proctor will yell "MISS" and the repetition will have to be repeated until done successfully. When the candidate completes the final set, "TIME" will be called out and the CPAT assessment will be completed.

Warnings and Failures

- If the candidate drops the pike pole and it hits the ground, a warning will be given; two warnings is a failure.
- If the candidate exceeds the 10:20 time limit, that is a failure.

> **Performance Point**
>
> In preparing for the CPAT it is important to learn and memorize all the "Warnings and Failures" for each event. Train with the events in mind and what each event is simulating in the real world of firefighting. During the CPAT you must stay mentally focused and aware, just as firefighters performing at emergency scenes.

CHAPTER SUMMARY

The Joint Labor Management Wellness-Fitness Initiative was developed by the IAFC and the IAFF to promote wellness and fitness among firefighters and to increase the quality of life for all fire department members. The WFI achieves its goal by focusing on five specific areas: medical evaluation, fitness testing and exercise, rehabilitation, behavioral health promotion, and data collection. A task force composed of IAFF and IAFC members, 10 fire departments, and a technical committee, was established to create an equal and fair assessment of the physical abilities of new firefighter candidates. The task force developed the CPAT, which consists of eight tasks done consecutively in a time not to exceed 10 minutes and 20 seconds. This evaluation promotes diversity, assesses the candidate's ability to perform on the fire ground, and helps fire departments to select quality individuals for their departments.

CHECK YOUR LEARNING

1. Who wrote and implemented the Joint Labor Management Wellness-Fitness Initiative (WFI)?
   a. Each fire department writes its own.
   b. The President of the Firefighter's Union, along with the Fire Chiefs of 10 fire departments.
   c. The International Association of Fire Chiefs (IAFC) and the International Association of Fire Fighters (IAFF).
   d. A special task force was developed.

2. One component of the WFI requires that all uniformed personnel have a yearly physical exam by a physician.
   a. True
   b. False

3. The goal of the WFI is to
   a. Decrease workers compensation claims.
   b. Promote wellness and fitness, improve the quality of life for all uniformed personnel, and prolong a quality career and life for all firefighters.
   c. Find the minority of the toughest and strongest men in the community.
   d. Discipline those that cannot keep up the high physical standards.

4. All of the following are part of the five topics of importance in the WFI *except*
   a. Data collection.
   b. Fitness testing and exercise.
   c. Body fat composition.
   d. Behavioral health promotion.

5. Which of the following constitutes a failure?
   a. Keeping the feet within the toe-box area during the Forcible Entry event.
   b. Performing 4 sets of 3 pushes and 5 pulls during the Ceiling Breach and Pull event.
   c. Placing the saws on the ground to re-grip during the Equipment Carry event.
   d. Running once the hose is draped across the body during the Hose Drag event.

6. Which of the following are *not* included in the CPAT guidelines?
   a. A hard hat with chin strap and gloves must be worn.
   b. Candidates must walk on the 85-ft transition between events.
   c. Candidates must wear a total 75-lb weight vest throughout the assessment.
   d. The wearing of jewelry or watches is not permitted.

7. Who was *not* on the technical committee of the CPAT task force?
   a. Attorneys.
   b. Line officers.
   c. Family members.
   d. Physicians.

8. The CPAT meets the validation criteria of all of the following *except*?
   a. U.S. Department of Labor (DOL)
   b. U.S. Department of Justice (DOJ)
   c. Canadian Human Rights Commission (CHRC)
   d. Firefighters Fair Labor Commission (FFLC)

9. Which task is *not* a part of the CPAT?
   a. Ventilation.
   b. Equipment Carry.
   c. Search.
   d. Ladder Raise and Extension.

10. The pass/fail time limit for the CPAT is
   a. 10 minutes.
   b. 8 minutes and 20 seconds.
   c. 12 minutes and 15 seconds.
   d. 10 minutes and 20 seconds.

References

1. International Association of Fire Fighters, AFL-CIO,CLC. (1999). *The Fire Service Joint Labor Management Wellness/Fitness Initiative.* Washington, DC: Author.
2. American Council on Exercise. (2003). *ACE Personal Trainer Manual (3rd ed.).* San Diego, CA: Author.

CHAPTER 2

THE CPAT: THE FOUR SUCCESS PRINCIPLES

LEARNING OBJECTIVES

Upon completion of this chapter the student will be able to:

- List and explain the four principles of success in completing the Candidate Physical Ability Test.
- List three reasons why goal setting is important.
- Define the SMART goal-setting technique.
- Define what a learning point is and give two examples.
- Define and give examples of active practice and overlearning.
- List five guidelines associated with feedback.

INTRODUCTION

Most successful physical skill training programs consist of four principles: goal setting, skill learning, practicing, and receiving feedback.[1] Whether you are preparing individually or as part of a group, these principles provide a framework by which you can acquire the knowledge, skills, and abilities necessary for successful completion of the Candidate Physical Ability Test (CPAT). Each principle is described in this chapter with guidelines and examples of how to implement them into your training schedule.

Figure 2-1 Goal setting will guide your efforts.

PRINCIPLE ONE: SETTING GOALS

"Is this the right way?" said Alice to the Cheshire cat.

"That depends a lot on where you want to go," said the cat.

"I don't know where I'm going," said Alice.

"Then it doesn't much matter which way you go," said the cat.

FROM LEWIS CARROLL'S MASTERPIECE, *ALICE IN WONDERLAND*

As the above quote demonstrates, successful training for the CPAT requires setting goals. **Goal setting** is the foundation on which all of your training and practice will be designed. It will guide what you do, when you will do it, and at what intensity **(Figure 2-1)**. When combined with the other training components, it will allow you to gauge the effectiveness of your training and when to make adjustments or take a rest day. As you accomplish your goals, you will train with confidence, vision, and enthusiasm. Train without setting goals and you will lack direction, purpose, mental focus, and most of all, results.

Why Goal Setting Is Important

The setting of goals affects performance in a dramatic way. An individual's behavior and intentions are regulated by the goals they have set.[2] You might have the necessary skills and be prepared physically, but it is your appraisal of what is to be done, how well you are prepared to do it, and whether you think it can or cannot be done, that affects the quality of your performance. When you clearly know what you want and are determined to reach it, your goals will harness that psychic energy and direct it toward the desired outcome.

Performance Point

Peak performers enjoy the experience of doing the work itself. Their strongest desires are for achievement (setting goals and reaching them) and self-actualization (being the best they can be). It is the setting of the goal itself and the working toward it that brings about increased performance.[3]

Goal Setting Guidelines

Step 1: Define your goals for the CPAT by using the **SMART** technique.

S–Specific. Specific goals lead to higher levels of achievement than generalized goals or the setting of no goals at all.

M–Measurable. You must be able to measure progress as you go along. If your goals are not measurable, you won't know where you stand on the road to your goal.

A–Attainable. You must believe that the goal is possible. You will not subconsciously commit to an unattainable goal. Setting goals above your past level of performance provides incentives and motivation to reach a bit further than last time.

R–Realistic. Set realistic and challenging goals. Realistic, challenging goals result in higher levels of performance than do no goals or generalized goals such as "do your best." If the goal is too easy, you will lose interest and commitment.

T–Time based. You must develop a timeline for each of your goals. This will allow you to know exactly where you should be at any specific point in time. List exact dates and times, and include a date for final completion of the goal.

Step 2: Goals should be set by you and written down. By setting and writing down your goals, you will subconsciously commit to them.

Step 3: Your initial goals should be simple and do-able. Accomplishing these small goals will build the foundation necessary for setting more challenging goals later on.

Step 4: Under each goal make a list of **objectives**. What are you going to do each day, each week to bring you closer to your goal? As you accomplish each of these objectives your confidence will grow.

Step 5: Write down how you will reward yourself when you achieve a goal. Draw up a goal/rewards contract with your significant other.

Step 6: Sign your written goal plan. You will then have committed yourself to take action.

Step 7: Evaluate your goals often. You will need to adjust any unmet goals.

> **Performance Point**
>
> Set your goals involving things over which you have control. Do not set goals over uncontrollable situations, such as weather conditions, judges, delays in testing, illnesses, course layouts, and minor injuries.

PRINCIPLE TWO: LEARNING THE EVENTS

Training for the CPAT without some idea of how successful candidates performed at each station would lead to guesswork, wasted effort, and frustration. A more effective way of learning the CPAT events is by observing successful candidates, reading about the techniques involved, and receiving proper instruction **(Figure 2-2)**.

Figure 2-2 Good instruction will help you learn the CPAT skills.

Why Learning Is Important

Through the process of observing others, we form ideas of how behaviors are performed and the effects they produce. When learning a physical skill such as the CPAT, observing how successful candidates perform on the course is the ideal method.

There are three different outcomes that result from your learning sessions:

1. **Positive outcome:** the learning situation results in better performance on the CPAT.
2. **Negative outcome:** the learning situation results in poorer performance on the CPAT.
3. **Neutral outcome:** the learning situation has no effect on CPAT performance. Use the following guidelines to maximize positive outcomes from learning situations.

Learning Guidelines

a. Maximize the similarity between your training situation and the CPAT. The CPAT is a physical skill, so learning by watching an actual candidate taking the test, whether live or on tape, is necessary for skill development.

b. You should watch for **learning points** in each event **(Figure 2-3)**. Learning points are key behaviors or techniques which lead to success in each event.

c. You should learn the basics first, adding the more complicated techniques later.

d. Learn from as many successful candidates as possible.

e. Try to view several examples of instances that do and do not represent the concepts being taught.

Figure 2-3 Watch for the learning points in each event.

Performance Point

To maximize positive outcome, models should be as similar to you as possible in age, sex, and capability. Women candidates should view techniques from women models if possible. It is difficult for a 6′4″ trainee to identify with a 5′6″ model.

PRINCIPLE THREE: PRACTICING THE EVENTS

Passing the CPAT requires repeated practice of the essential skills involved in each of the eight stations (as shown in **Figure 2-4**).

There are two key components of practice: **active practice** and **overlearning**.[3]

Active practice provides you with as much experience as possible with the actual or simulated CPAT course. Overlearning entails practicing far beyond the point where you have mastered each task.

Why Practice Is Important

Repeating the essential movements involved in the CPAT will provide your body with the internal cues that regulate motor performance. As you continue to practice, internal cues leading to errors are eliminated and internal cues associated with smooth and precise performance are retained.

Practice Guidelines

a. Active practice requires practicing on the actual CPAT course or a close simulation. The program design chapter has ideas for

Figure 2-4 Practice with perfect form.

simulating each of the CPAT course stations if you do not have access to the actual course.

b. Overlearning makes your performance more **reflexive** or automatic **(Figure 2-5)**. This increases your ability to maintain the quality of your performance when taking the actual CPAT.

c. The ability to mentally focus during your practice sessions is important. Module 4 has guidelines on developing focusing skills.

d. Practice the skills under a wide range of conditions. This will prepare you for unexpected events at the testing site.

> **Performance Point**
>
> In the early stages of training, when you are trying to master the skills, a complex task like the CPAT should be broken down into parts and learned progressively, one section at a time.

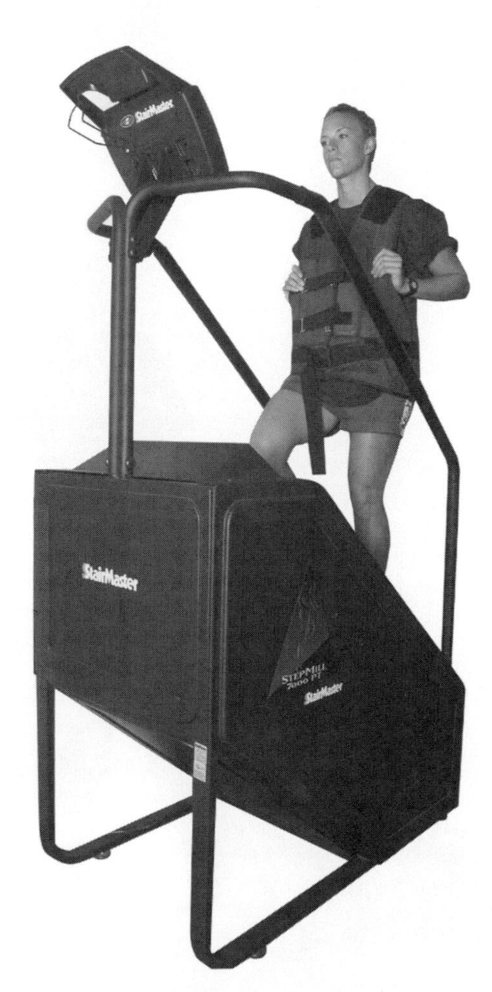

Figure 2-5 Overlearning makes your performance automatic.

PRINCIPLE FOUR: FEEDBACK

Imagine competing in a sporting event without keeping score or attempting to learn a new skill and receiving no feedback. Now imagine training for the CPAT and having no idea of your times for each event, whether your technique is efficient, or whether the nutritional plan you started is effective. You would be frustrated and your motivation would drop. When learning a new physical skill like the CPAT, feedback or knowledge of results is critical for both learning and motivation **(Figure 2-6)**.

Why Feedback Is Important

Having clear goals is not enough. You must feel, moment by moment, that you are on your way toward reaching those goals. Feedback serves three purposes:

1. It conveys information to you about your performance.
2. It keeps your motivation high.
3. It allows you to set or modify your goals.

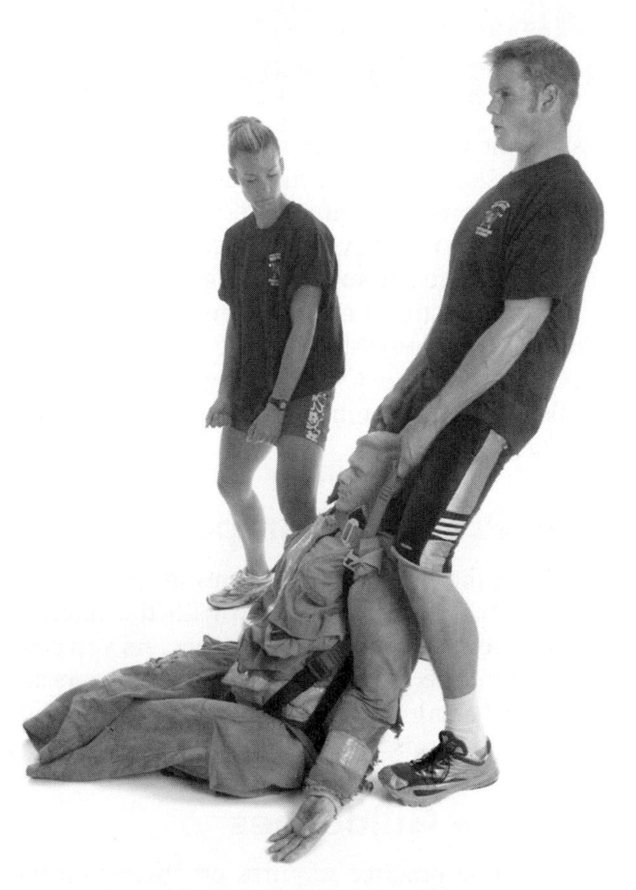

Figure 2-6 Receiving feedback is crucial to your success.

Feedback Guidelines

1. Feedback is neither good nor bad in itself. It is simply information that can be used to monitor and adjust your performance as needed.
2. The feedback from movements of the body allows you to make ongoing adjustments to keep on track or to get back on it. You know when movements have the right "feel" or your performance was "smooth." Known as **kinesthetic awareness**, this feel for one's performance is a key source of feedback when training for the CPAT **(Figure 2-7)**.
3. Ideally, feedback should be provided as soon as possible after your performance. The CPAT coach, training staff, and other firefighters can provide this immediate feedback.
4. Results from physical fitness tests will provide you with feedback on your physical strength and endurance capability.
5. Feedback from performance measures such as your actual time on the CPAT or a simulated course should be broken down into segments: time at each station, time between each station, and total time for the entire course. This will allow you to pinpoint areas where improvement is needed.

Figure 2-7 Developing kinesthetic awareness will help your performance become more automatic.

Performance Point

It is difficult to keep a positive attitude in the face of criticism or negative feedback. It is easy to lose confidence by focusing on errors and perceiving that you are failing. Yet, it is what you decide to do with incoming information that has the biggest influence on your state of mind. What should you do when it seems that things are not going well? Michael Jordan said in an advertisement, "I've failed over and over again in my life, and that is why I succeed." This quote by Jordan is also a reminder that challenge is the key to successful experiences. How the mind reacts is at least as important as the actual error itself.

Case Study: Using the Four Success Principles

Case 1

Angie, a firefighter candidate, set six goals in preparation for the CPAT. She set the following goals after going through initial fitness testing and a trial run-through on the CPAT course.

Goal 1—Decrease her body fat percentage from 26% to 20%.

Objective:
Lose 0.5% body fat per week for 12 weeks. She will increase her meal frequency from 2 meals per day to 6 meals per day. Angie will increase her meal frequency by 1 per day for each week for a period of 4 weeks until she is up to 6 meals per day. Increase her water intake to 1 gallon daily.

Goal 2—Increase her upper and lower body strength.

Objective:
Consult with a certified personal trainer for help in developing and planning a training schedule.

Goal 3—Decrease her 1.5-mile run time from 18:32 to 12:00.

Objectives:
Complete two high-intensity cardiovascular workouts per week.
Complete two low-intensity cardiovascular workouts per week.

Goal 4—Learn the CPAT course and develop a strategy for each event.

Objectives:
View a tape of a successful CPAT candidate.
Interview three firefighters, including at least one female who have completed the CPAT. Find out what worked for them and what techniques and tips they can relay to her. Find out what mental focus techniques worked for these candidates.
Enroll in a CPAT preparation course offered by the local community college.

Goal 5—Decrease her time for the CPAT from 16:11 to under 10:00.

Objectives:
Break down her total cóurse time into time at each event and time between each event.
Analyze her technique at each station and pinpoint where improvements could be made.
Develop a mental focus for each of the practice stations.

Goal 6—Receive feedback on her training sessions.

Objectives:
Have biometric testing done initially and every 4 weeks. This will provide feedback on the effectiveness of her strength and cardiovascular training goals.
She will time her CPAT trial run sessions: total time, time at each station, and time between each station.
She will videotape her practice sessions. This will allow her to analyze her technique.

Case 2

Bill, a former Marine, set three goals in preparation for the CPAT.

Goal 1—Break the course record of 8:31.

Objectives:
Run up the stairs at the local high school 3 times per week.
Complete three high-intensity weight-training sessions each week.

Goal 2—Increase his protein intake.

Objectives:
Drink three protein shakes per day.
Include 8 oz. of protein at each of his regular meals.

Goal 3—Become familiar with the CPAT course.

Objectives:
Take a trial run on the CPAT course.

Case Study Questions

1. How has Angie incorporated the four success principles in her training plan?
2. How well has Bill incorporated the four success principles into his training plan?
3. Are the goals and objectives Angie set effective in helping her pass the CPAT?
4. What other goals could Bill have set to help him prepare for the CPAT?

CHAPTER SUMMARY

Successful preparation for the CPAT consists of four key principles: goal setting, learning the skills, practicing the skills, and receiving feedback.

Setting goals using the SMART technique (specific, measurable, attainable, realistic, and time orientated) will direct and guide your training.

Learning and practice goals should be reevaluated and adjusted according to feedback received from practice sessions and trial runs on the CPAT.

To be effective, goals should be written down and signed, and be rewarded when accomplished.

When learning the CPAT it is important to note key behaviors or learning points which lead to success in each event. Maximize your learning by studying candidates who are similar to you in sex, body size, etc.

Active practice involves training as much as possible on the actual or simulated CPAT course. Overlearning involves practicing the skills far beyond the point of mastery. To maximize your training, practice should be done under a variety of weather and course conditions, and you should develop a mental focus for each of the stations.

Feedback provides information on your performance, keeps your motivation high, and allows you to set or modify goals. A key source of feedback is kinesthetic awareness which is generated from movements of the body while practicing. Feedback from CPAT trials should be received as soon as possible after your practice sessions and broken down into time at each station, time between each station, etc.

CHECK YOUR LEARNING

1. Goal setting principles include
- **a.** Writing down your goals.
- **b.** Listing objectives for each major goal.
- **c.** Writing broad, unspecific goals.
- **d.** Evaluating your goals often.
- **e.** a, b, and d.

2. The SMART goal setting technique includes
- **a.** Time orientated goals.
- **b.** Extra ordinary goals.
- **c.** Do your best goals.
- **d.** Nonspecific goals.

3. The most effective way to learn the CPAT is
- **a.** Try it yourself.
- **b.** Interview several successful candidates.
- **c.** Watch a video of a successful candidate.
- **d.** Attend a training class on the CPAT.
- **e.** All of the above.

4. What should you do to ensure a positive learning experience?
- **a.** Learn the basics first.
- **b.** Maximize similarity between your training and the CPAT.
- **c.** Watch for key behaviors and learning points.
- **d.** All of the above.

5. Overlearning a skill means
- **a.** Mentally focusing while practicing.
- **b.** Practicing until you are exhausted.
- **c.** Practicing under a wide range of conditions.
- **d.** Practicing a skill beyond mastery.

6. Active practice involves
- **a.** Learning each station beyond mastery.
- **b.** Mentally focusing while practicing.
- **c.** Receiving feedback while you practice.
- **d.** Increasing practice time at a station.

7. Kinesthetic awareness is
- **a.** Neither good nor bad in itself.
- **b.** Feedback from movements of the body.
- **c.** Watching an actual candidate take the test.
- **d.** A learning point.
- **e.** Monitoring your heart rate while practicing.

8. Feedback serves what purpose?
- **a.** Allows you to set or modify your goals.
- **b.** Allows you to practice beyond fatigue.
- **c.** Keeps your motivation high.
- **d.** Conveys information about your performance.
- **e.** a, c, and d.

9. A learning point is
  a. A diagram of the CPAT course.
  b. An interview of a successful CPAT candidate.
  c. A key behavior in successful performance.
  d. Focusing on a skill.

10. Feedback should be
  a. Provided at least 1 week after your performance.
  b. Provided for each segment of your practice sessions.
  c. Can be discarded if provided by a nonfirefighter.
  d. Should not be provided by your coach or teammates.
  e. b and d.

References

1. K. Wexley, G. Latham, *Developing and Training Human Resources in Organizations.* (Santa Monica, CA: Goodyear, 1981).
2. E.A. Locke, et al., "Goal setting and task performance: 1969–1980," *Psychological Bulletin*, vol. 90, pp. 125–152, 1981.
3. Wexley, Latham, *Developing and Training Human Resources in Organizations.*

CHAPTER 3

LEARNING THE SKILLS

LEARNING OBJECTIVES

Upon completion of this chapter the student will be able to:

- Explain the value of biometric testing in designing a Candidate Physical Ability Test (CPAT) training program.
- List the steps involved in raising one's lactate threshold.
- Understand the role that complex carbohydrates play in completion of the CPAT.
- List the six major nutrients and explain their role in energy production.
- Describe the importance of ingesting protein and carbohydrates for the post-workout meal.
- Explain the use and value of a heart rate monitor when training for the CPAT.
- List the five training principles and explain their role in designing a CPAT training program.
- Explain the role that mental training plays in maximizing CPAT performance.
- Understand the concept of "centering" and explain why it is critical to successful completion of the CPAT.

MODULE 1
Biometric Testing

Figure M1-1 Testing your cardiovascular system.

INTRODUCTION

The creator of Famous Amos Cookies was asked, "Where do you start when opening a new business?" Famous Amos replied, "Start from where you are at." This first learning module will help you determine starting points or baseline measurements for your training exercises. It will determine your strengths and weaknesses, help you set goals, and improve your Candidate Physical Ability Test (CPAT) skills. The feedback you receive from ongoing testing sessions will enable you to evaluate your goals and re-set them if necessary.

WHY BIOMETRIC TESTING IS IMPORTANT

There are several reasons why initial and ongoing testing is beneficial and necessary. Initial testing will allow you to compare your capabilities to the specific demands of the CPAT. You will be able to customize a training program that progresses in a safe, injury-free, productive manner. Initial testing will allow you to set realistic and motivating goals.

Testing can screen for possible health risks and **musculoskeletal** problems.

Feedback provided by re-testing will help you evaluate your goals and re-set them if necessary. Re-testing will also tell you if you are overtrained and need to scale back your training.

GUIDELINES FOR BIOMETRIC TESTING

The following tests are accurate predictors of the skills and abilities required to pass the CPAT. They are easy to do and do not require a lot of specialized or expensive equipment.

1. Resting heart rate (RHR). Your resting heart rate indicates your basic fitness level. The more conditioned you are, the less effort and fewer beats per minute it takes your heart to pump blood to your body. To determine your resting heart rate

Figure M1-2 Using a heart rate monitor will provide training feedback.

put on your **heart rate monitor** or manually count the beats for 1 minute **(Figure M1-2)**. Use your fingers to count the pulse at your wrist or at the carotid artery at the side of the neck. Do this when you wake up before you get out of bed in the morning for 3 days in a row and then average the readings. If your RHR rises, it could indicate that you are overtrained. See Chapter 5 for directions for treating overtraining.

2. Balance. Balance is the foundation of all movement and it is crucial for all CPAT events. Put on a 50-lb weight vest and perform the following tests. Record your times for each of these tests and record them on the performance planning chart.

Balance progression:

a. Stand on one leg for 10 seconds. Then repeat on the other leg.
b. On the dome side of a **BOSU Balance Trainer**, stand on one leg **(Figure M1-3)**. Repeat standing on the other leg. If you do not have access to a BOSU, use an inner tube or similar object that creates an unstable surface. Most health clubs have BOSUs, Dyna Discs, or other balance training types of equipment.
c. Turn the BOSU over and stand, then squat on the flat side **(Figure M1-4)**. Performing the exercise with weights will increase the difficulty and effectiveness of this movement.
d. Stand on the floor and balance on one leg with your eyes closed.

3. Body composition. Excess body fat has a direct effect on your CPAT performance. Losing excess body fat will increase your quickness and endurance, and reduce the likelihood of injury and joint pain. There are several methods to measure body composition. To ensure validity, use the same technique and technician for your repeat testing: **hydrostatic or underwater weighing**, is accurate but expensive and time consuming. Most

Figure M1-3 One-leg balance.

Figure M1-4 Squat on the flat side of a BOSU, with weights.

universities are equipped to do hydrostatic weighing. Bioelectrical impedance requires you to satisfy several conditions, including your **hydration level** (amount of water in the body), time since your last exercise session, and time since last meal. This technique is not recommended for lean individuals as it will overestimate your body fat level. **Skin fold measurement** is the most common method used. This technique requires an experienced technician to get an accurate reading.

Dual-energy x-ray scan (DEXA) is accurate but expensive. It is the most intriguing method because it shows the amount and distribution of fat by location.

4. Flexibility. Flexibility is the range of motion of a joint. A lack of flexibility will inhibit smooth performance on the CPAT. The sit-and-reach method is the most common **(Figure M1-5)**. You can also estimate your flexibility by sitting on the floor with your legs extended and attempt to touch your toes. Do not bend your knees. If you can touch your toes or go beyond, your flexibility is acceptable. If you can't touch your toes, then you need extra stretching work.

5. Push-ups. Upper body strength and endurance is measured by the 1-minute push-up test. Start in the up position then lower your torso down to your partner's fist. Then push yourself back up to the top position. Record the number of push-ups completed in 1 minute.

6. Sit-ups. The ability to transfer force generated by your lower body into the execution of each event is essential for passing the CPAT. Test your abdominal strength by performing as many sit-ups as possible in 1 minute. Have someone hold your feet and count each repetition. Sit-ups are used for testing purposes only. Crunches and other core body exercises are used for training purposes (see Module 3).

Figure M1-5 Testing your flexibility.

7. Bench press. The bench press is used to assess upper body strength. **One repetition maximum (1RM)** is the maximum amount of weight that can be lifted for one repetition. However, lifting the maximum amount of weight you can may lead to injury. Using a spotter for safety, place enough weight on the bar that allows you to perform 10 repetitions (10RM). To calculate your 1RM, divide the 10RM weight by 0.75. Divide this by your body weight to determine body weight/bench press ratio.

8. Lactate threshold. Your **lactate threshold (LT)** is the point where you start using carbohydrates instead of fats as a primary energy source. Module 3 has more details on the fuel sources used in exercise intensity. By determining your LT you will be able to calculate your specific heart-rate training zones. Knowing these zones and how to intersperse them in a training program will enable you to improve your CPAT performance.

To determine your LT, map a 1.5-mile course. Warm up, then run the route at the fastest pace you can sustain. Record your highest heart rate by either using a heart rate monitor or manually counting your pulse. Your LT is the approximately the highest heart rate you can sustain for the 1.5 course.

9. CPAT trial run. Going through the actual CPAT course at the start of your training will provide you with feedback on your ability to pass the test. Be sure to have someone track your times at each event and times between events.

Performance Point

Going through all of the biometric tests is the ideal situation. However, if you do not have access to or cannot perform all the tests, then do as many as possible. The more feedback you have, the better.

PERFORMANCE PLANNING CHART

Biometric Testing

Use this chart to record your biometric testing results. This will provide you with the feedback necessary to develop a training program for the CPAT.

BIOMETRIC TESTING RESULTS

	Standard F(Female) M(Male)	Baseline Date:	Retest Date:	Retest Date:	Retest Date:	Retest Date:	Retest Date:
Resting heart rate	<60						
Balance	All tests completed						
Body composition	F: <24% M: <17%						
Flexibility	F: >17 M: >15						
Push ups	F: >26 M: >29						
Sit ups	F: >39 M: <44						
1RM bench press	F: >0.7 M: >1.0						
Lactate threshhold HR	Varies						
CPAT trial run	<10:20						

List your goals for the biometric testing results:

Goal 1 ____________________ **Completion date:** ____________

Objectives: ____________________

____

Feedback source: ____________________

Goal 2 ____________________ **Completion date:** ____________

Objectives: ____________________

____

Feedback source: ____________________

Goal 3 ____________________ **Completion date:** ____________

Objectives: ____________________

____

Feedback source: ____________________

Figure M2-1 Good nutritional planning is vital to your success.

MODULE 2
Nutritional Concepts

INTRODUCTION

"Never eat more than you can lift," said Miss Piggy. If Miss Piggy were to train for the CPAT, she would find that there is more involved in good nutrition than your ability to lift it. This module will outline the principles and guidelines of high-performance nutrition **(Figure M2-1)**. It will show you what to eat and how to time your meals for maximum CPAT performance. You will have the energy to train at the high-intensity level necessary to improve your lactate threshold, endurance, and explosive power. These same principles are used by Olympic competitors, triathletes, and other elite athletes.

You will also decrease the chances of developing overtraining symptoms and injuries, and your mental function will be sharp and focused on each event.

WHY NUTRITION IS IMPORTANT

Within a year's time every cell in your body will die off and new ones will replace them. These new cells are constantly developing; forming connective tissue, muscle tissue, skin, blood, hormones, and enzymes.

The foods you eat are the building blocks of this redevelopment process.

Eating quality foods will give you the energy to train at the high-intensity level necessary to pass the CPAT. Consuming poor or "junk" foods will reduce your ability to train effectively and will decrease your chances of passing the CPAT.

There are six major nutrients: carbohydrates, proteins, fats, vitamins, minerals, and water.

Carbohydrates are the main energy source for the body and the most important nutrient when exercising. They are easily converted to **glycogen**, which the body uses for energy production. There are two types of carbohydrates: **simple carbohydrates** are easily digested and quickly restore glycogen stores in the muscles. Simple carbohydrates such as energy drinks and gels can supply a quick source of energy. **Complex carbohydrates** are slower to digest and provide energy for longer periods. Among all the nutrients, they are the most powerful in affecting your energy levels **(Figure M2-2)**.

Performance Point

Low carbohydrate diets do not have a place in the dietary needs of an active, healthy, athletic person. Complex carbohydrates are essential for peak muscle performance and brain function. Low carbohydrate diets can cause decreased performance, reduced mental function, and muscle tissue breakdown.

Protein is used to repair tissue damage, build the immune system, and help the body to recover from exercise. Proteins are the building blocks for all tissue, enzymes, and hormones that control our movements and metabolism.

You should ingest 2.0 grams per kg (2.2 lbs) of body weight per day.

Good protein sources include chicken breast, turkey breast, egg whites, fish, lean meats, nonfat dairy products, and protein powders.

Fats carry and store fat-soluble vitamins, construct cell membranes, and play a role in the

Figure M2-2 Complex carbohydrates will give you explosive power.

Figure M2-3 Water is the best drink.

production of testosterone and estrogen. There are two types of naturally occurring fats. **Saturated fats** are solid at room temperature and come from animal sources. Examples of saturated fats include beef, lamb, bacon, pork, and lard products. Consuming large amounts of saturated fats contributes to many disease processes. **Unsaturated fats** tend to be liquid or semi-liquid at room temperature. Unsaturated fats are divided into polyunsaturated and monounsaturated. These are from plant sources and are necessary for good health and peak performance. Examples include flax oil/seeds, avocado, walnuts, almonds, and salmon. **Trans fats** (partially hydrogenated ingredients) raise the levels of bad cholesterol (LDLs) and increase the risk of heart disease. For more examples of each nutrient, see **Table M2-1**.

Vitamins and minerals are not used for energy but they are needed for your body to convert macronutrients (carbohydrates, protein, and fat) to energy. **Vitamins** assist in the production of red blood cells, help to liberate energy from fuel stores, fight against free radicals, and help in tissue repair. **Minerals** help in tissue repair, play a role in maintaining a regular heart rhythm, and are necessary for muscle contraction.

Water is necessary for optimum endurance, maintaining body temperature, removing waste products, and metabolism **(Figure M2-3)**. When you lose as little as 2% of your body weight due to dehydration your performance will decrease by 10% to 15%.

> **Performance Point**
>
> Eat for performance and not for losing weight. If you have been dieting, increase your intake of complex carbohydrates for a couple of weeks and see how your body responds. You might be surprised to find that you do not gain weight and see your performance improve. In the long run, making the right choices will pay off with big dividends (see **Figure M2-4**).

GUIDELINES FOR NUTRITION

To help sustain high-intensity training throughout the day (Figure M2-4):

- Eat five or six meals each day. Eating every 2 to 3 hours will elevate your metabolism, keep your energy high, and promote overall body leanness.

Table M2-1. High Performance Food Chart

SERVING SIZE GUIDELINES			
Serving size: 1 serving = ½ cup, 1 medium, or slice Whole grains (rice, pastas, breads oats, cereals): 5–7 servings per day Vegetables: 3–5 servings Fruits: 2–4 servings per day			
COMPLEX CARBOHYDRATES (LONG TERM ENERGY/PRE-WORKOUT)			
Brown rice Oatmeal Quinoa Barley Peaches Pears Beans Yam Sweet potato Grapefruit	Fat free yogurt Orange Apple Banana Dates Blueberries Melon Figs Strawberries Kiwi	Grapes Asparagus Tomato Spinach Broccoli Green beans Sprouted grains Kale Onions Cranberries	Red peppers Green peppers Mushrooms Whole-grain bread Whole-grain pasta Brussels sprouts Black-eyed peas Cabbage Tomato paste
SIMPLE CARBOHYDRATES (POST WORKOUT)			
Potato Orange juice Raisins Honey	Natural granola Ripe bananas Carrots Beans	100% Fruit juices Sports drinks Whole grain cereal Whole wheat bagel	Whole grain pretzels Whole grain crackers
PROTEINS			
Chicken breast Turkey breast Salmon Light tuna Egg whites Buffalo	Cottage cheese Soy Hummus Brown rice & beans Buckwheat Lentils	Beans Tofu Fat-free yogurt Peanuts Skim milk Corn & beans	Pasta and bean soup Low-fat string cheese Natural peanut butter Corn tortillas & beans
GOOD FATS			
Salmon Light tuna Sardines Almonds	Walnuts Flax seeds/oil Peanuts Olive oil	Canola oil Avocado Natural peanut butter	Fish oil capsules Sunflower seeds
DRINKS & SEASONINGS			
Water Green tea	Vegetable juice Oregano	Salsa Picante sauce	Cinnamon

Figure M2-4 Frequent meals will provide you with energy for high-intensity training.

Figure M2-5 Track your nutrition by using a food journal.

- Always eat breakfast because it improves physical and mental performance. Skipping breakfast is associated with a higher risk of obesity.
- Drink 8 to 12 glasses of water a day. Avoid drinking a lot of fluids when eating solid food, as this will slow down your digestion.
- Always read food labels. Each gram of fat has 9 calories, each gram of protein and carbohydrate has 4 calories. Limit or avoid all foods with hydrogenated oils, saturated fat, and chemical-laden ingredients.
- Plan ahead what you are going to eat each day. Use a cooler and some portable, reusable food storage containers for healthy snacks. To avoid feeling deprived, pick one day of the week to eat whatever you wish.
- Include all six of the major nutrients in your diet (see **Table M2-1**). Organic is best. Cut out alcohol or limit yourself to only one drink a day.
- Track your nutrition by keeping a food journal **(Figure M2-5)**. This valuable tool will provide you with feedback on what you eat, portion size, and how much water you drink a day.
- Proper nutrition produces high performance. Supplements are not recommended because of the side effects associated with them.
- Losing or gaining weight depends on how many calories you expend versus how many you consume. Do not crash diet—you will lose muscle mass and weaken yourself.
- Strive for the 90% rule. Eat healthy 6 days per week consuming high-performance foods. Pick one day where you eat whatever you want (nachos, pizza, cake or a piece of pie, high-caloric dinners, etc.). This will keep you from feeling deprived.

Before exercise:

- Eat a pre-workout meal about 1.5 hours prior to exercising. Eat complex carbohydrates and a little protein, such as a whole wheat bagel with a thin layer of peanut butter on it.

- Stay away from fats, as they metabolize slowly. Be sure to have a pre-event meal on CPAT test day. An empty stomach is likely to cause you to fatigue more quickly and run out of energy.
- Drink at least 8 oz of water 15 minutes before exercise.

During exercise:

- During your workout, drink 4 to 6 oz of water every 10 to 15 minutes. Drink before you feel thirsty.
- If it is excessively hot or humid consume more water. If your workout lasts longer than 1 hour use a sports drink or gel to replenish glycogen stores.

After exercise:

- Consume 300 to 400 calories of protein and simple carbohydrates within 30 minutes after the workout. This is important because the body needs to replenish glycogen stores and start repairing tissue.
- Replace each pound of fluid lost with 16 to 24 oz of water.

Performance Point

Keeping a food journal (see **Figure M2-5**) will enable you to track how your nutritional program is affecting your training. You will be able to modify your nutritional program if needed and gauge how many calories you consume a day. There is a sample food journal in Appendix C.

Goal Setting

Set your nutritional goals below. Be sure to include a completion date and objectives.

PRE- AND POST-EXERCISE IDEAS FOR COMBINING CARBS AND PROTEIN
1. Energy bar and 8-oz sports drink
2. Two slices whole grain toast and 2 tablespoons peanut butter
3. Orange and ½ cup lowfat cottage cheese
4. One cup cooked oatmeal with ¾ cup raisins
5. One cup yogurt and ¼ cup granola
6. Two-egg omelet with 1 cup fresh vegetables, 1 whole wheat muffin
7. String cheese and 1 oz of pretzels
8. Quarter cup nuts and a medium apple
9. Hardboiled egg and ½ whole wheat bagel
10. Whole wheat pita and ½ cup canned tuna
11. Quarter cup soy nuts and 1 banana
12. Quarter cup sunflower seeds and 1 cup orange juice
13. Three oz chicken breast and 1 cup cooked brown rice

PERFORMANCE PLANNING CHART

Nutrition

CUSTOMIZING YOUR CPAT NUTRITIONAL PLAN

Match your nutritional plan with the training phase you are in (see Chapter 5). The ratio of carbohydrates, proteins, and fats, will be dictated by the intensity and type of exercise you are doing. Figure your daily calorie needs, then calculate the amount of carbohydrates, proteins, and fats you will need.

There are four training phases:

Phase 1 is a skill development phase and your energy needs will be moderate.

Phase 2 will increase training intensity. CPAT training drills will last longer and strength training will be emphasized. Your energy requirements will increase.

Phases 3 and 4 will focus on developing power. You will need more energy from increased protein and carbohydrates.

Training Phase	Intensity Level	Calories Needed Per Pound M: male F: female bw: body weight	Total Daily Calories Needed	Carbohydrate Calories/ Grams	Protein Calories/ Grams	Fat Calories/ Grams
Phase 1	Moderate	M: 20 × bw F: 20 × bw		× 65% =	× 15% =	× 20% =
Phase 2	High	M: 23 × bw F: 20 × bw		× 65% =	× 20% =	× 15% =
Phase 3	Very High	M: 24 × bw F: 21 × bw		× 65% =	× 20% =	× 15% =
Phase 4	High	M: 23 × bw F: 21 × bw		× 65% =	× 20% =	× 15% =

Goal Setting

Set your nutritional goals below. Be sure to include a completion date and objectives.

Goal 1 ______________________ Completion date: __________

Objectives: ______________________

____

Feedback source: ______________________

Goal 2 ______________________ Completion date: __________

Objectives: ______________________

____

Feedback source: ______________________

MODULE 3
Fitness Training Principles

Figure M3-1 Training the core will improve your CPAT performance.

INTRODUCTION

Learning and implementing sound training principles is necessary for your success in the CPAT **(Figure M3-1)**. Training programs which include a variety of techniques have shown to produce better results than programs that focus on a single sport or activity.[1] This module will outline an effective multidimensional approach to training that will prepare you for the CPAT. Given jobs, family, school, and life in general, most CPAT candidates cannot afford huge blocks of time in the gym. The recommended training programs in this manual are tailored to fit the "normal" candidate's busy lifestyle.

TRAINING PRINCIPLES

1. Specificity. The body responds and adapts to the type of physical demands under which it is placed. To pass the CPAT, you must condition yourself to the actual CPAT events or close replicas. The more similar your training is to the demands of the CPAT, the more likely your performance will be successful.[2]

2. Overload. The body will adapt to an increase in physical demands placed on it. For continued improvement you have to gradually make the program harder with intermittent periods of rest and recovery built in. This will allow you to avoid overtraining and injury, yet provide enough stress to allow adaptation to occur. Chapter 5 describes training plans that will help you to accomplish this goal.

3. Reversibility. The body's fitness level will decline if you discontinue your exercise program. Without appropriate exercise, muscles become smaller and weaker, the cardiovascular system less efficient, and flexibility levels decrease. To keep your CPAT skills sharp you must train with sufficient intensity and frequency.

4. Individuality. The capacity of each individual to handle a given workload is unique. If you and a friend do exactly the same training in precisely the same way, you probably will not get the same results. Some individuals are "fast responders" while others are "slow responders." This is probably genetic. Women generally need more recovery time than men, and older individuals require longer recovery time than younger individuals.

5. Warm up and cool down. Always warm up for 5 to 10 minutes prior to beginning an exercise session. Use light running, treadmill walking, biking, etc. A thorough warm-up will increase blood flow to joints and muscles, increasing freedom of movement, and reducing the risk of injury. A proper warm-up will make exercise feel more comfortable and place less stress on the body.

Cool down after the workout by stretching for 5 minutes, using relaxed breathing. This is the ideal way to return your body to normal after exercise.

6. Determine your starting points. Use biometric testing results from Module 1 to determine your starting points for each of the components. This will also allow you to gauge your progress and make adjustments if necessary.

THE CPAT TRAINING PROGRAM

The CPAT training program consists of five components. Each component is described below in detail with reasons why it is important for your success in the CPAT. Use the results of your biometric testing sessions from Module 1 to determine the starting points for each of the training components.

TRAINING PHASES

Your training will be divided into four phases that are designed to gradually increase strength and skill development. Chapter 5 has several comprehensive training schedules with guidelines on how you can integrate these phases into an overall training plan.

Component 1: Strength Training

There is no doubt that **strength training (Figure M3-2)** will increase your performance. Research has shown that athletes enjoy an increased "time to exhaustion" following a strength training program.[2] The main purpose of strength training for the CPAT is to increase your ability to produce the power that is required for success in each of the events. Appropriate strength training will also reduce injuries, decrease fatigue, and enhance your cardiovascular capacity.[3]

If you are combining strength training with other components, perform the strength training first. This will allow you to maximize strength benefits.

Your starting weights are determined by your biometric testing results from Module 1.

Figure M3-2 Squatting develops strength.

DEFINITIONS AND GUIDELINES

- A **repetition** (or **rep**) is a successful completion of an exercise movement.
- A **set** is a group of repetitions.
- Exercises involving barbells are labeled **bb**; exercises involving dumbbells are labeled **db**.
- Think progression, adding more weight when the resistance seems easier. Never sacrifice proper execution for more weight.
- If your resistance training exercises from start to finish are taking longer than 1 hour to complete, you are not training hard enough or are taking too much time between exercises. The idea is to get in and out of the gym effectively and efficiently.
- Try to make the exercise more difficult by slowing down and eliminating as much momentum as possible. Harder exercises, not easier, are what stimulate muscle growth.
- Keep the number of strength training exercises low and focus on major muscle groups: chest, back, glutes, and legs. This will ensure maximum strength for the "power" movements which more closely simulate the actions performed in the CPAT.
- Perform the exercises in the order listed. The multi-joint exercises (squats, lunges, rows, bench presses, and partial dead lifts) are performed before the single joint exercises.

Phase 1—Strength Training Exercises

Do 2 sets of 12 repetitions. Lift slowly through the full range of motion—about 2 seconds to lift, 4 seconds to lower the weight.

Workout 1A
1. squats
2. bb or db bench press
3. front lat pull down
4. lying leg curl
5. db curl

Workout 1B
1. leg press
2. db shoulder press
3. db row
4. seated calf raise
5. incline db press

Phase 2—Strength Training Exercises

Perform 2 sets of 10 repetitions, with a gradual increase in weight load, maintaining proper execution. This will train the muscles to tolerate higher blood lactate levels.

Workout 2A

1. squats
2. bb bench press or db bench press
3. front lat pulldown
4. lying leg curl
5. db curl

Workout 2B

1. leg press
2. db shoulder press
3. partial dead lift
4. db row
5. seated calf raise
6. incline db press
7. tricep pushdown

Phase 3—Strength Training Exercises

This phase is divided into moderate (endurance) exercises and heavy (power) exercises. When lifting heavier weights in the power sets, use 4 to 6 seconds to lift and 4 seconds to lower the weight. When doing endurance sets, use 2 seconds to lift and 2 seconds to lower the weight. Use 2 sets of 12 repetitions for the endurance sets, and 3 sets of 8–6–4 reps for the power sets. As the reps decrease, the weight load increases.

Workout 3A

1. squats or leg press (power)
2. bb bench press or db bench press (power)
3. lat pulldown (endurance)
4. lying leg curl (power)
5. db curl (endurance)

Workout 3B

1. close grip press (power)
2. db shoulder press (endurance)
3. partial deadlift (power)
4. db row (endurance)
5. seated calf raise (endurance)
6. incline db press (endurance)

Phase 4—Strength Training Exercises

Use 2 sets of 12 repetitions for the endurance sets, and 2 sets of 6 repetitions for the power sets. Remember to think control when performing the endurance sets, 2 seconds to lift, 2 seconds to lower the weight.

Workout 4A

1. db row (power)
2. walking lunge (endurance)
3. bb or db bench press
4. bb row or db row
5. partial dead lift (power)
6. close grip press (power)

Workout 4B

1. squat or leg press (power)
2. Cable pushdown (endurance)
3. db press (endurance)
4. incline db press (power)
5. seated row (endurance)
6. reverse grip pull-down (endurance)
7. ez bar curl (endurance)

Performance Point

Some muscle soreness should be expected after exercise, especially if you're working at an intense level. Muscle soreness is caused by the microscopic breakdown of muscle tissue from intense exercise. The subsequent repair helps you get stronger as the muscle adapts to the increased demand.

Component 2: Cardiovascular Training

To complete the CPAT, you must train both the **aerobic** and **anaerobic** energy systems. The aerobic system **(Figure M3-3)** uses oxygen primarily for low-intensity training and physical activities lasting more than 3 minutes. The aerobic system builds the foundation for the high-intensity anaerobic system. The anaerobic system's primary energy source is glucose which is developed from carbohydrates. This system produces a high amount of energy and causes the accumulation of waste products in the muscles and blood, such as **lactic acid**. During physical activity, both systems are active at specific times depending on the intensity of the program.

To improve your performance, it is necessary to train at levels that will increase your **lactate threshold (LT)**, the point when you transition from aerobic to anaerobic energy systems. When you raise LT, you can produce more power at a comfortable heart rate, and that makes you a better performer in every event. Training intensity is divided into five zones **(Table M3-1)**. Determine your heart rate for each of the zones by applying your lactate threshold heart rate (from Module 2) to the following chart. This is best accomplished by using a **heart rate monitor (Figure M3-4)**. If you don't have a heart rate monitor, you can subjectively estimate your training zone level by using the **rating of perceived exertion (RPE)** method. The RPE is a subjective assessment of the intensity of your exercise. Use

Figure M3-3 Step-ups train the entire lower body.

the feeling description (column three) to estimate your training zone (column one).

> **Performance Point**
>
> For the longest time we focused on finding your **target heart rate zone**—the range of heart rates optimal for improving your fitness. However, there isn't a single target zone that works best for training. There are actually several zones you should spend various amounts of time training within. In addition, the formula commonly used to determine the target zone (220 minus your age) can be wildly inaccurate for any given person. Using the lactate threshold is a completely individual physiological response to exercise.

Phase 1—Cardiovascular Exercises

You can use any exercise mode you wish (treadmill, bike, running, elliptical, etc.). Using the Step Mill as much as possible will lead to an overlearning effect.

Figure M3-4 Using a heart-rate monitor will provide training feedback.

Start each workout with a warm-up and end each workout with a cool-down.

Workout 1A—Warm up for 5 minutes at a low intensity. Get your heart rate up to zone 1 and maintain it for 15 minutes. Finish with a 5-minute cool-down/stretch. For weeks 3 and 4, get your heart rate up to zone 2.

Workout 1B—Warm up for 5 minutes at a low intensity. Get your heart rate up to zone 2 and maintain it for 25 minutes. Finish with a 5-minute cool-down/stretch.

Phase 2—Cardiovascular Exercises

Workout 2A—Warm up for 5 minutes at a low intensity, then keep your heart rate 3 to 5 beats below your LT for 10 minutes (zone 3). Recover for 3 minutes then repeat two more times.

Workout 2B—Warm up for 5 minutes at a low intensity. Get your heart rate up to zone 1 and maintain it for 35 minutes.

Phase 3—Cardiovascular Exercises

Workout 3A

Option 1: Pyramid (35 minutes)

1. Warm up for 5 minutes at a low intensity.
2. Get your heart rate up to zone 1 and maintain it for 5 minutes.
3. Minutes 6–10 up to zone 2.
4. Minutes 11–15 up to zone 3.
5. Minutes 16–18 up to zone 4.
6. Minutes 19–30 down to zone 2.
7. Do a 5-minute cool-down, followed by stretching.

Table M3-1. Training Intensity Zones.

Zone Description	Heart Rate Range	Rating of Perceived Exertion (RPE) Feeling
Zone 5 Anaerobic Speed & Power (Extremely Hard)	110 percent of LT to MHR	Extremely difficult and uncomfortable. Intense desire to slow or stop. Cannot be sustained long-term.
Zone 4 Anaerobic Endurance (Very Hard)	100 to 110 percent of LT	Breathing becomes heavy, difficult and uncomfortable. You may experience muscle burn.
Lactate Threshold		Start to feel "burn" in muscles.
Zone 3 Aerobic Endurance (Hard)	90 to 100 percent of LT	Breathing becomes noticeable, but not too difficult. Conversation becomes limited.
Zone 2 Aerobic Development (Medium)	70 to 90 percent of LT	Breathing deepens a bit. Conversation is more difficult. Urge to go faster.
Zone 1 Warm up (Easy)	60 to 70 percent of LT	Comfortable to talk and breathe through nose.

Option 2: Scott drills (perform the following exercises in order for 10 minutes)
1. Step Mill climb with weight vest for 3 minutes.
2. Step Mill backwards climb for 2 minutes.
3. Treadmill run/walk at 8% incline for 3 minutes.
4. Versa Climber or Elliptical for 2 minutes.

Repeat steps 1–4 two more times.

Option 3: Sprint drills

Minute	**Intensity Level**
1–5	Warm up (low intensity)
6	Zone 3
7	Zone 2
8	Sprint for 10 seconds up to zone 5, then down to zone 2 for 20 seconds; repeat.
9–11	Zone 2
12	Sprint for 10 seconds up to zone 5, then down to zone 2 for 20 seconds; repeat.
13–15	Zone 2
16	Sprint for 10 seconds up to zone 5, then down to zone 2 for 20 seconds; repeat.
17–20	Cool down, followed by stretching.

Workout 3B—Warm up for 5 minutes at a low intensity. Get your heart rate up to zone 2 and maintain it for 25 minutes. Finish with a 5-minute cool-down/stretch.

Phase 4—Cardiovascular Exercises

Workout 4A

Option 1: Up and down LT intervals—Warm up, then move to your LT heart rate (zone 4) and hold that intensity for 5 minutes. Push it to about 3 to 5 beats above LT for 1 to 2 minutes, then drop it back down to LT. Continue for a total of three cycles, or about 18 to 20 minutes.

Option 2: LT Tolerance Intervals—Warm up then increase your effort to 5 beats above your LT heart rate (zone 4). Hold it there for 2 minutes. Increase your effort to zone 5 for 30 seconds. Reduce your effort for 90 seconds, just long enough so you feel partially recovered, but not quite ready to go again. Repeat three times.

Workout 4B—Warm up for 5 minutes at a low intensity. Get your heart rate up to zone 1 level and maintain it for 35 minutes.

Component 3: CPAT Specific Drills

The fitness principle of specificity states that in order to excel in a given area you must train that area. CPAT specific drills involve training in the actual CPAT events or close simulations **(Figure M3-5)**. There are many types of drills and exercises that you can perform which are designed specifically to replicate each CPAT event. **Table M3-2** lists ideal and alternate training drills for each CPAT event. See Modules 5 to 12 for specific descriptions of each technique.

You will also need to incorporate transition training into your CPAT specific drills. Transitions are the 85 feet you must travel between each event. How you perform on these seven transitions can make or break your performance.

Transition times should also be used to prepare yourself physically and mentally for the next event.

Figure M3-5 Mastering the events requires training on them or a close simulation.

CPAT Specific Drills Chart

Circuit Station	Ideal Exercise	Alternate Exercise
1	Step Mill	Stairmaster, stair or step climb
2	Hose drag	Tire rope drag
3	Saw, carry	db carry, paint can carry
4	Ladder raise/extn	Ladder raise, cable pulldown
5	Hammer swing	Ball and rope swing, cable swing
6	Tunnel trail run	Mat tunnel
7	Dummy drag	Sack or duffel bag drag
8	Cable one-arm pulldowns	db or knapsack pulldowns
	Cable one-arm raises	db raises, knapsack raises

Table M3-2. CPAT-Specific Drills Time Chart.

Training Phase	Workout	Step Mill Time	Times per Stations 2-8	Transition Time per Station	Number of Cycles	Weighted Vest lbs. (Step Mill)
One	1A	2 minutes	30 seconds	30 seconds	1	none
	1B	2 minutes	60 seconds	30 seconds	1	none
Two	2A	3 minutes	60 seconds	30 seconds	2	50 (50)
	2B	3 minutes	90 seconds	30 seconds	2	50 (75)
Three	3A	4 minutes	120 seconds	30 seconds	2	50 (75)
	3B	4 minutes	120 seconds	30 seconds	2	50 (75)
Four	4A	5 minutes	180 seconds	30 seconds	2	50 (75)
	4B	4 minutes	180 seconds	30 seconds	2	50 (75)

GUIDELINES

- Set up the stations in a circuit arrangement. Start with station 1, then follow the steps in order.
- Completion of all eight stations and transitions equals one cycle.
- Follow the time per station and transition time per station guidelines listed on **Table M3-2** to determine the time spent at each station.
- During the transition periods, try to lower your heart rate by breathing deep and relaxing all muscles not used in walking.

> **Performance Point**
>
> Even though an evaluator will provide you with directions, you should think through your transitions ahead of time. During each transition, try to lower your heart rate and prepare yourself mentally for the next event. You will be walking 85 feet—relax all muscles except the ones involved in walking. Prepare yourself mentally for each transition by checking out the course ahead of time as CPAT courses vary in their setup.

Figure M3-6 Lunging develops balance skills.

Component 4: Functional Skills Training

Functional skills training develops key CPAT skills that are not trained in the other four components **(Figure M3-6)**. Included are training for lower back, abdominal, balance, and reaction time. Functional skill training enhances all the other components because the forces applied by your arms and legs must pass through the core of your body. If the core body is weak, much of the force is dissipated and this will result in poor performance (see **Tables M3-3, M3-4, M3-5** and **M3-6**).

Figure M3-7 Stretching after your workout will increase flexibility.

Component 5: Flexibility

Flexibility is defined as the capacity to move freely in every intended direction **(Figure M3-7)**. When movement is compromised by tight muscle groups, performance diminishes and exposes risk of injury. Stretching increases physical efficiency as a flexible joint requires less energy to move through its range of motion. A consistent and effective program of stretching can prevent and reduce the risk of injury and low back pain. At the end of each workout, cool down by using **static stretching** to adequately recover and reduce later muscle soreness. Static stretching involves going into a stretch position and holding it at your point of limitation for 15 to 30 seconds. Avoid bouncing as this will tighten muscles.

Flexibility Exercises for All Phases

Duration: Hold each stretch 20 seconds.
Repetitions: Perform each stretch one time.

Lower Body Stretches

- Seated hamstring stretch: Standing calf stretch (straight leg)
- Seated inner thigh stretch: Standing calf stretch (bent leg)

Table M3-3. Phase One Functional Skill Exercises–1A.

Exercise	Equipment Used	Exercise Description	Sets & Reps
Basic crunch	Fit Ball	Lying face up on fit ball	2 sets, 10 repetitions
Angled crunch	Fit Ball	Lying face up on fit ball	2 sets, 10 repetitions
Walking lunges	Fit Ball	Hold fit ball over head while lunging	2 sets, 10 repetitions

Table M3-4. Phase Two Functional Skill Exercises–2A.

Exercise	Equipment Used	Exercise Description	Sets & Reps
Basic crunch	Fit Ball	Lying face up on fit ball	2 sets, 10 repetitions
Angled crunch	Fit Ball	Lying face up on fit ball	2 sets, 10 repetitions
Walking lunges	Medicine ball	Hold medicine ball extended from chest level and rotate towards outside of front leg when lunging	2 sets, 10 repetitions
BOSU standing & squatting balance	BOSU	2 legged standing and squatting on dome side of BOSU	1 set, 10 repetitions

Table M3-5. Phase Three Functional Skill Exercises–3A.

Exercise	Equipment Used	Exercise Description	Sets & Reps
Basic crunch	Fit Ball & medicine ball	Lying face up on BOSU or fit ball, hold medicine ball at chest level when performing exercise	2 sets, 10 repetitions
Angled crunch	Fit Ball & medicine ball	Same as above	2 sets, 10 repetitions
Walking lunges	Fit ball	Hold fit ball overhead when performing exercise	2 sets, 10 repetitions
BOSU standing & squatting balance	BOSU	2 legged standing and squatting on dome side of BOSU	2 sets, 10 repetitions

Table M3-6. Phase Four Functional Skill Exercises–4A.

Exercise	Equipment Used	Exercise Description	Sets & Reps
Basic crunch	Fit Ball	Lying face up on fit ball	2 sets, 10 repetitions
Angled crunch	Fit Ball & medicine ball	Lying face up on fit ball, extend medicine ball arms length and then rotate side to side	2 sets, 10 repetitions
Walking lunges	BOSU	2 legged standing and squatting on flat side of BOSU	1 set, 10 repetitions add: hand weights when mastered
BOSU standing & squatting balance	Dyna disc	One-legged squat on Dyna Disc	1 set, 10 repetitions for each leg—add weighted vest when mastered

- Lying hip stretch: Hip flexor stretch
- Standing quadriceps stretch: Groin straddle

Upper Body Stretches

- Lying lower back stretch: Shoulders (cross arms in front of chest)
- Lying spinal twist stretch: Triceps (behind neck stretch)
- Shoulders stretch: Shoulders straight arms behind back

Performance Point

For a long time it was recommended that individuals should stretch before a competition or before their workouts. It was thought that stretching reduces the incidence of injury and pulled muscles. This recommendation was given mostly out of intuition and was not based on any hard data. Now it is thought that warming up by increasing blood flow to muscles is a better way to prepare for exercise and prevent injury. According to experts, warming up increases blood flow, speeds up nerve impulses, and assists in the removal of waste products. All stretching should be done at the end of each workout.

References

1. R.C. Hickson, et al., "Potential for Strength and Endurance Training to Amplify Endurance Performance," *Journal of Applied Physiology*, vol. 65, pp. 2285-2290, 1988.
2. Mel Siff, et al., *Super Training: Special Strength Training for Sporting Excellence* (Johannesburg, South Africa: School of Mechanical Engineering, University of the Witwatersrand, 1993).
3. M.H. Stone, et al., "Health and Performance-related Potential of Resistance Training," *Sports Medicine*, vol. 11, pp. 210-231, 1991.

PERFORMANCE PLANNING CHART

Strength Training

Name: **Training Phase:**

DATE																		
Exercise	Wt/ rep	Wt/ rep	Wt/ rep	Wt/ rep	Wt/ rep	Wt/ rep	Wt/ rep	Wt/ rep	Wt/ rep	Wt/ rep	Wt/ rep	Wt/ rep	Wt/ rep	Wt/ rep	Wt/ rep	Wt/ rep	Wt/ rep	Wt/ rep

PERFORMANCE PLANNING CHART

CPAT Specific Drills

Name: **Training Phase:**

CPAT Event	Drill	Duration	Heart Rate Zone	Transition Time	Notes
Step Mill					
Hose Drag & Carry					
Equipment Carry					
Ladder Raise & Extension					
Forcible Entry					
Search					
Rescue					
Ceiling Breach & Pull					

PERFORMANCE PLANNING CHART

Functional Skills

Name: **Training Phase:**

Date	Exercises	Equipment Used	Duration	Sets/Repetitions	Notes

PERFORMANCE PLANNING CHART

Cardiovascular Training

Name: **Training Phase:**

Date	Training Mode	Training Technique	Duration	Heart Rate Zone	Notes

MODULE 4
Mental Training for the CPAT

Figure M4-1 Mental training allows you to apply your skills.

INTRODUCTION

You have spent hours training on CPAT-related drills and on the actual CPAT course. You have trained your cardiovascular, core body, and muscular systems. Your nutritional plan is sound and all your fitness tests have improved.

Besides physical skills, mental skills also need to be learned, trained, and mastered. Most great athletes acknowledge that training the mind as well as the body is necessary to achieve success.

WHY MENTAL SKILLS ARE IMPORTANT

Final results are determined by how you use mental training techniques to apply your skills. Your mind "guides" your body when performing. It allows your body to apply the skills you have developed. Mental and muscle memory interact, and you can train them together to maximize your performance. When you strengthen your mental skills, your confidence is strengthened and your commitment is enhanced.

PRINCIPLES OF MENTAL TRAINING

Visualization—Visualization trains you to mentally experience each event as if you were living it. Perform daily sessions where you visualize yourself completing the CPAT with perfection. By imagining yourself successfully completing each event, your subconscious mind will "tell" your conscious mind to perform in that manner.

When practicing, picture yourself doing the skills, then do each skill, letting it unfold naturally, trusting your body to do what its been trained to do. When you actually take the test, it will be a finalization of something you have done before.

Focusing—Focusing allows you to maximize your performance by excluding all irrelevant thoughts and emotions. A positive focus connects you totally with your performance **(Figure M4-1)**. Determine what focus will work for you and use it to direct your mind when performing each event.

Staying focused throughout the entire CPAT is a challenge. You will also need to develop a re-focusing strategy to respond swiftly to lapses in focus. Use the re-focusing strategy to shift your attention away from worry on to performing the next skill.

Self-Talk—Do the things you say to yourself make you feel like a failure, or confident and strong? All positive thoughts are constructive. All negative thoughts are destructive. The constructive force of positive self-talk will lead you to success in practice **(Figure M4-2)** and in the CPAT. Examples of positive self-talk are **mantras** (a phrase such as "quick feet" or "strong body"). You can use positive self-talk as your focus points also. For example, when performing the stair climb event, use the mantra "strong legs." Remember, only you can determine what focus points and self-talk will work for you.

Relaxation—Stress makes your mind hurry and your muscles tense up. When you are feeling stressed, slow everything down. Use breathing techniques to relax and consciously loosen tight muscle groups. Control tension in specific muscles by tensing and then relaxing a muscle group. Repeat this process throughout the entire body. If you tense up when testing for the CPAT, breathe deeply three or four times and focus on a positive thought. The ability to relax and maintain proper form results in less energy lost, increased confidence, and faster times **(Figure M4-3)**.

Self-Control—Are you willing to accept the input and advice of your coach, other candidates, instructor, or significant other?

Worry diminishes performance. Do you waste energy worrying about things beyond your control? Are you disciplined enough to use re-focusing techniques to regain lost focus when necessary? Do you use relaxation techniques to calm yourself in stressful situations?

Centering—Centering is the breathing and focusing process that you go through to position yourself for optimal performance. Being centered allows you to transfer all your power into your movements. Your center of mass is a spot located just below and behind your navel. When you are centered, your knees are slightly bent and your weight is evenly distributed between your feet **(Figure M4-4)**. If a boxer raises his center of mass by locking his knees before throwing a punch, he cannot get his weight or power behind the punch. If the boxer begins with a low center of mass (knees bent) and then transfers his weight by raising the center of mass as he jabs, he has the full weight of his body behind the punch. Shifting your center of mass through your body gives power to movements.

Figure M4-2 Positive self-talk will help you avoid choking.

Figure M4-3 Relax by using deep breathing.

Figure M4-4 Proper centering position.

Performance Point

Learning to center yourself and then transferring the center of mass through your body and into the event you are practicing is the key to performing well. This skill is developed by using relaxed breathing and focus points (see **Figure M4-4**). At first, this process will be conscious and mechanical. Thought processes often interfere with your ability to feel centered. With quality practice, the process will become quicker and automatic and you will be able to maintain freedom from distractions for longer periods. Examples of positive self-talk/mantra for centering:

Loose	Controlled	Relaxed
Confident	Solid	Powerful
Balanced	Fluid	Strong
Calm	Light	Tranquil
Energetic	Peaceful	Effortless
Easy	Commanding	Clear
Smooth	Focused	

Find words that are most powerfully associated with performing at your best. Which words seem to have the strongest emotional impact on you? Those two words will act as triggers to help you establish a feeling of being centered.

GUIDELINES FOR MENTAL TRAINING

1. Set goals for mental training as you would for physical training. The attainment of goals will increase your self-confidence and open the door to optimal performance.

2. During each event, use strong, positive focus points. This will help you avoid distractions. Recall your best past performances and the feelings and focus associated with them. Find out what works for you. Focus on what you should be doing. Do not focus on what you shouldn't be doing.

3. Develop a re-focusing strategy. When you get distracted, use simple reminders, images, or mantras to regain your focus. Re-focus with positive self-talk, such as "strong core" or "strong legs."

4. Make self-talk simple and positive. Remind yourself of your best performances, good recent practices, and your ability to perform well.

5. Prior to each practice or test, visualize yourself performing with flawless technique. Analyze your performance and pinpoint any flaws. Then visualize yourself overcoming those flaws.

6. Use relaxation techniques and positive self-talk to shift your focus away from worry. If your focus is centered on something other than worry, you cannot be worrying at the same time.

7. Your relaxation responses must be well learned. Start by practicing under low-stress conditions, then under medium stress, and finally under high-stress conditions.

8. Use breathing and focusing techniques to center yourself before each event. As you practice, shift your center of mass toward the event when you are performing.

9. Use self-talk and visualization during your days off. This will train your subconscious and will transfer to your actual performance.

10. Develop a support system. Your family, friends, coach, significant other, etc. will help you push through hard workouts.

11. Self-control helps you develop the mental skills required to perform your best. These skills are developed long before the day of the test through preparation and experiences that teach you to maintain or regain control over your mental state.

Performance Point

No matter how much an individual has prepared for the CPAT, the possibility of choking or making mistakes is a reality. Following the training plans presented in this manual will reduce this likelihood and help you to recover from mistakes. The difference between a successful and unsuccessful candidate is preparation—physically and mentally.

PERFORMANCE PLANNING CHART

Mental Training

Use this chart to help you develop a mental training strategy for your practice sessions and the CPAT.

ANALYZING YOUR MENTAL PREPARATION

Determine your readiness for practice.

Check

________ Have you set goals for each practice session?
________ Did your preparation help you to meet your goals?
________ Did you visualize yourself executing each practice session with perfect technique?
________ Are you using positive self-talk?
________ Do you perform relaxation techniques daily?
________ Do you have a focus point for each event?
________ Can you maintain your best focus when faced with distractions?
________ Do you seek out and accept feedback from your instructors, coaches, and fellow firefighters?
________ Do you control any negative emotions?
________ Do you have the discipline to follow your training schedule?
________ Do you have a support system developed?
________ Do you practice with a "relaxed focus" and allow your body to execute the skills you have trained it to do?
________ Do you sustain your focus throughout the duration of each practice session?
________ Do you center your self before each practice event?
________ Are you using "trigger" words to start the centering process?

Determine what is your best performance focus and self-talk for each CPAT event. Then determine what your response will be if you lose that focus.

Event	Focus Point/Self-Talk	Re-focusing Response
1. Stair Climb		
2. Hose Drag		
3. Equipment Carry		
4. Ladder Raise & Extension		
5. Forcible Entry		
6. Search		
7. Rescue		
8. Ceiling Breach & Pull		

Case Study: Determining Nutritional Requirements

Firefighter candidates Angie and Bill set the following nutritional goals:

Goal 1—Develop a nutritional plan for their CPAT training.

Objective: Determine their daily total caloric needs.

Objective: Break down their daily calories into carbohydrate, protein, and fat percentages.

Goal 2—Develop a feedback mechanism for their nutritional program.

Objective: Use the nutrition log in Appendix C to track their progress.

Case 1

Angie: Body weight (bw) = 130 lbs

Training Phase	Calorie calculation: Daily calories needed	Daily calories needed from: Carbs	Protein	Fat
1	20 × bw = 2600	65% = 1690	15% = 390	20% = 520
2	20 × bw = 2600	65% = 1690	20% = 520	15% = 390
3	21 × bw = 2730	65% = 1775	20% = 546	15% = 409
4	21 × bw = 2730	65% = 1775	20% = 546	15% = 409

Case 2

Bill: Body weight (bw) = 200 lbs

Training Phase	Calorie calculation: Daily calories needed	Daily calories needed from: Carbs	Protein	Fat
1	20 × bw = 4000	65% = 2600	15% = 600	20% = 800
2	20 × bw = 4000	65% = 2600	20% = 800	15% = 600
3	21 × bw = 4200	65% = 2730	20% = 840	15% = 630
4	21 × bw = 4200	65% = 2730	20% = 840	15% = 630

CHAPTER SUMMARY

A successful CPAT training program is composed of initial and ongoing biometric testing, well-conceived training principles, effective mental focusing abilities, and a solid nutritional plan.

Initial biometric testing enables candidates to compare their capabilities to the specific demands of the CPAT. Starting points for all training components are determined and program goals are set. Re-testing will provide feedback on program effectiveness and enable candidates to evaluate and re-set goals, if necessary.

Nutritional planning is necessary to meet the intense training demands of the CPAT. Each candidate's nutritional plan should be designed to meet the changes in program intensity, prevent overtraining, and reduce the occurrence of injuries.

Proper meal timing, meal frequency, and water intake will increase energy and aid recovery. Maintaining a food journal will enable candidates to track the effectiveness of their nutritional program.

The recommended CPAT training program consists of five components: cardiovascular, strength training, CPAT specific skills, functional skills, and flexibility/stretching. These components will develop the general conditioning and specific skills necessary to pass the CPAT. Training schedules are divided into four phases of increasing intensity, which are designed to gradually increase strength and skill development.

Mental training will allow the body to apply the skills developed during practice sessions. This will strengthen a candidate's confidence and enhance commitment. The most crucial mental skill, focusing, is used to eliminate outside influences and concentrate on the event being performed.

All the mental skills are used to perfect the ability to center, which allows the transfer of lower body force into the production of power.

CHECK YOUR LEARNING

1. Centering involves
   - **a.** Focusing on the center of your body.
   - **b.** Eliminating outside distractions when you are practicing.
   - **c.** Transferring the force generated from your lower body into your performance.
   - **d.** Focusing on the center of the event.

2. Strengthening your mental skills will result in
   - **a.** Your mistakes being minimized.
   - **b.** Your strength increasing.
   - **c.** Your confidence increasing.
   - **d.** Your commitment being enhanced.
   - **e.** a, c, and d.

3. Visualization involves
   - **a.** Checking out the CPAT course ahead of time.
   - **b.** Seeing yourself executing your movements with perfect form.
   - **c.** Watching successful candidates take the CPAT.
   - **d.** Watching unsuccessful candidates take the CPAT.
   - **e.** a and b.

4. The most effective way to increase your explosive power is to
   - **a.** Raise your lactate threshold.
   - **b.** Master your ability to center yourself.
   - **c.** Develop a focus point for each event.
   - **d.** Run on the transitions between events.

5. The overload principle states that
   - **a.** Skills must be maintained or they will degenerate.
   - **b.** People will progress at different rates.
   - **c.** Pushing the body slightly past its strength level.
   - **d.** You must train an activity to master it.
   - **e.** Determining your starting points before exercising.

6. Which energy system does the CPAT use?
   - **a.** The aerobic system.
   - **b.** The oxygen system.
   - **c.** The anaerobic system.
   - **d.** The RPE system.
   - **e.** a and c.

7. Which nutrient provides the most energy for the CPAT?
- **a.** Simple carbohydrates.
- **b.** Complex carbohydrates.
- **c.** Fats.
- **d.** Proteins.
- **e.** Fiber.

8. Which of the following represents the ideal meal frequency and nutrient combination?
- **a.** 3 meals per day with protein at every meal.
- **b.** 1 low-carbohydrate meal per day.
- **c.** 5–6 meals per day with high protein at every meal.
- **d.** 8 high-fat meals daily.
- **e.** 6 meals daily with balanced protein and carbohydrates at each meal.

9. Complex carbohydrates consist of the following
- **a.** Brown rice, whole grain pasta, and oatmeal.
- **b.** White rice, cottage cheese, and whole grain pasta.
- **c.** Oatmeal, orange juice, and green beans.
- **d.** Potato, honey, and tomato.
- **e.** Skim milk, brown rice, and beans.

10. Biometric testing will
- **a.** Allow you to customize your training program.
- **b.** Reveal nutritional deficiencies.
- **c.** Provide you with starting points for each exercise.
- **d.** Measure your blood glucose levels.
- **e.** a and c.

11. Lactate threshold is
- **a.** The point your body transfers from aerobic to anaerobic energy systems.
- **b.** Your ability to produce power.
- **c.** Your ability to burn fat for extended periods.
- **d.** Your ability to produce speed.
- **e.** The point at which lactate decreases in the blood.

LEARNING THE EVENTS

LEARNING OBJECTIVES

Upon completion of this chapter the student will be able to:

- List and explain the physical demands of each of the eight CPAT events.
- Define the importance of kinesthetic awareness in completing the Step Mill and Search events.
- Explain how a centered position will enhance the transfer of power in completing each of the events.
- Demonstrate the most effective way of pulling the dummy in the Rescue event.
- Demonstrate the proper way to "lock the rope" for the Ladder Raise and Extension event.
- List the steps involved in the Equipment Carry event.
- Explain the importance of mental focus when completing the transfers between events.

MODULE 5
Stair Climb

Figure M5-1 Correct form: hands are away from side rails.

KEY STATS

Average success time:	3:00 minutes
Average transition time:	23 seconds
Cardiovascular demand:	High
Skills and abilities needed:	Balance, lower body strength and endurance
	Kinesthetic awareness of motion involved in stair climbing

TECHNIQUE AND STRATEGY

Many candidates struggle with the Stair Climb event. In fact, most people who fail the CPAT fail on this station. You step at a moderate 60 steps per minute for 3 minutes. However, you are carrying a 75-lb weight vest on your back. Your heart rate will quickly elevate to anaerobic levels and stay there for 3 minutes and 20 seconds. You will need to have good lower body, abdominal, and low-back strength to keep your body upright and support the additional weight.

You are not allowed to touch the side rails with your hands, except momentarily for balance. The ideal position for your hands is to either hold on to the weight vest in front of your body or to rest your arms at your sides **(Figure M5-1)**. Either of these positions will keep you centered and minimize the stress on your body.

Keep your body in proper alignment by keeping your head up and focusing on a point in front of you. If you look down at the steps you might become disorientated.

Performance Point

Listen to the proctor count down time at the end of the test. When the test is over, immediately dismount the machine so the proctor can remove the two 12.5-lb weights from your shoulders. If you delay in dismounting the Step Mill, it can cost you precious seconds.

CPAT SPECIFIC EXERCISES FOR THE STAIR CLIMB

Ideal Training Exercise

The best exercise is the actual Step Mill. Start by practicing walking on the Step Mill and learning to balance. Use the manual program to become familiar with the step height and pace of the machine. Remember, you are not allowed to hold on to the hand rails during the CPAT **(Figure M5-2)**, so your goal should be to step without holding on. Try stepping for at least 3 minutes and 20 seconds (the length of the test actual test). Once you have mastered the stepping cadence and balance, start stepping with a weight vest. Gradually increase the amount of weight for the vest until you reach 75 lbs. Because of the stress that 75 lbs places on your joints and back, it is not recommended that you increase the weight beyond this level. To produce an overlearning effect, you must also increase the amount of time on the Step Mill. Once you have achieved the 3:20 mark, increase your time to about 7 minutes. You will find the actual test easier if you accomplish this. Listen to your body. If your joints and back start to bother you, remove the added weight after 5 minutes and continue stepping until you reach 7 minutes.

Figure M5-2 Incorrect form: holding on to the side rails.

Alternate Training Exercises

If you do not have access to a Step Mill, use a StairMaster climber, a step platform used in a step class, or the first step of a staircase. The step should be at least 8 inches in height so you will need two of the riser blocks under each side of the step platform. Practice by stepping up with the right foot, then up with the left, down with the right foot, then down with the left foot. Make sure you place the entire foot on the step when stepping up **(Figures M5-3, M5-4)**. If part of your foot hangs off the step, you will place stress on the Achilles tendon. Because the lead foot and leg perform the majority of the work, make sure you alternate which foot leads or goes first. Alternating the lead foot will prevent you from over using one side of the body. You should strive to look straight ahead and keep your posture upright when stepping.

Use the same training sequence as was outlined for the Step Mill. After mastering the stepping cadence and balance, start stepping with a weight vest or backpack. Start gradually with 20 lbs, then 30 and 40, gradually working toward 75 lbs. Because of the stress that 75 lbs places on your

Figure M5-3 Correct stepping form.

Figure M5-4 Incorrect form: half of foot is off step.

joints and back, it is not recommended that you increase the weight beyond this level. Similar to the Step Mill training program, your goal should be to train far longer than the actual test. This will produce an overlearning effect, and will make the actual test easier. Once you have achieved 3 minutes and 20 seconds (the length of the test) gradually increase your time to about 7 minutes.

Listen to your body. If your joints and back start to bother you, remove the added weight after 5 minutes and continue stepping until you reach 7 minutes.

> **Performance Point**
> Effective strength, cardiovascular, and mental focusing training are necessary for mastering the Step Mill event. Lack any of these key components and your chances of experiencing premature fatigue or losing your balance will increase. If you lose your balance while stepping, push yourself away from the rail by using the back of your hand. Grabbing the rail more than twice will result in disqualification.

MENTAL FOCUS FOR THE STAIR CLIMB

Examples of mental focus for the Step Mill:

- Breaking the total time of 3 minutes down into 15 or 20 second manageable segments. Take one step at a time.
- See yourself going strong all the way through, stepping off at the finish knowing that you could have done more.
- Try not to fight or beat the Step Mill—experience it.

Successful candidates have a mental focus and plan for each event. Only you can determine what mental focus will work for you (see Chapter 3—Module 4 on mental training).

SKILL DEVELOPMENTAL EXERCISES FOR THE STAIR CLIMB

Refer to Appendix A for illustrations and performance tips for the following exercises **(Figures M5-5, M5-6)**:

Lower Body Strength & Endurance	Balance	Anaerobic Development
Squats	BOSU balance sequence **(Figure M5-5)**	Interval drills
Leg press	DynaDisc balance	Sprint drills
Lunges	Walking backwards on the Step Mill **(Figure M5-6)**	Scott drills
One-legged squats		
Bench step-ups with weight (10 reps with each leg)		Up and down LT intervals

Figure M5-5 Balance training is essential for the Step Mill.

Figure M5-6 Walking backwards on the Step Mill will train the back of your legs.

PERFORMANCE PLANNING CHART

Event 1: Stair Climb

GOALS

- Using the SMART technique, list your learning and practice goals for the Stair Climb. What challenges are you facing, and what specific actions do you need to do to meet these challenges? For each goal, list a completion date, objectives, and a feedback source.
- Evaluate each goal after completion. If necessary, set new goals to keep progressing.

LEARNING

Learning Points for the Stair Climb

- Keep your head up and look straight ahead while stepping.
- Keep your body in proper alignment: chest out, shoulders back.
- What other learning points will lead to success in the Stair Climb?

____

____

Learning Goals

List how you will learn to do the Stair Climb.

Goal 1 ________________________________ **Completion date:** ______________

Objectives: ____

____

Feedback source: ____

Learning outcome: ____ **positive** ____ **neutral** ____ **negative**

Goal 2 ________________________________ **Completion date:** ______________

Objectives: ____

____

Feedback source: ____

Learning outcome: ____ **positive** ____ **neutral** ____ **negative**

Continued

PERFORMANCE PLANNING CHART

Event 1: Stair Climb—Cont'd

PRACTICE

What mental focus will you use while practicing for the Stair Climb?

____

____

Practice Goals

When and where will you practice? Detail how you will overlearn and use active practice in your training sessions for the Step Mill.

Goal 1 ____________________ **Completion date:** __________

Objectives: ____

____

Feedback source: ____

Practice technique: _____ **Active practice** _____ **Overlearning**

Goal 2 ____________________ **Completion date:** __________

Objectives: ____

____

Feedback source: ____

Practice technique: _____ **Active practice** _____ **Overlearning**

Goal 3 ____________________ **Completion date:** __________

Objectives: ____

____

Feedback source: ____

Practice technique: _____ **Active practice** _____ **Overlearning**

MODULE 6
Hose Drag

Figure M6-1 Stay in the boxed area when pulling the hose.

KEY STATS

Average success time:	28 seconds
Average transition time:	19 seconds
Cardiovascular demand:	Moderate
Muscles involved:	Upper body strength, lower body strength
Skills and abilities needed:	Balance; kinesthetic awareness of the movements involved in carrying and pulling a hose from a kneeling position Grip strength

TECHNIQUE AND STRATEGY

You have just finished the most demanding cardiovascular event of the CPAT: the Step Mill. On the transition to the Hose Drag event **(Figure M6-1)**, try to lower your pulse as much as possible. Focus your mind and visualize yourself performing the event. When you reach the hose, pick up the nozzle, place it on your shoulder, and move toward the drum. Travel a couple feet beyond the drum—this will minimize the drag on the hose when you make the right turn. Once you have reached the box, drop down on one knee and start to pull the hose. Center yourself and keep your back in proper alignment. This will make your pulling motion more efficient.

Remember to keep your knee on the ground and stay within the box.

Visualize yourself pulling the hose with a steady, rhythmic hand-over-hand motion.

The goal of this event is to lower your pulse as much as possible. Also, try to be within a few seconds of the average success time of 28 seconds.

Performance Point

Some candidates choose to run once they pick up the hose. How much time does this really save? You are coming off the most demanding cardiovascular event (Step Mill) and running with the hose will prevent you from lowering your heart rate. In the long term, lowering your pulse is the most important consideration.

CPAT SPECIFIC EXERCISES FOR THE HOSE DRAG

Ideal Training Exercise

Place a drum 75 ft away from the starting point and mark a 5 × 7 ft box at a 90-degree angle 25 feet from the drum. Use a 100-ft length fire hose which is marked 8 × 50 ft from the nozzle. Drape the hose over your shoulder at the 8 ft or less mark. Drag the hose to the left side of the barrel **(Figure M6-2)**, go 2 ft beyond it then turn right and go 25 ft to the marked box. Drop down to one knee and pull the hose until the 50-ft mark is in the box **(Figure M6-3)**. You can overlearn this event by practicing with a heavier hose.

Alternate Training Exercise

Attach at least 50 ft of rope to a tire or cement block. Mark the rope 8 ft and 50 ft from the front end. Drape the rope with not more than 8 ft (the first marker) over your shoulder. Drag the resistance 75 ft to a drum or similar object. Make a 90-degree turn and drag the rope another 25 ft to a box marked on the ground. Drop to one knee and pull the hose to the second mark (50-ft mark) in the box. Start with light resistance and concentrate on technique. When you have developed good kinesthetic awareness for the pulling motion, increase the resistance by adding another tire or small weight. Do not go heavier than 50 lbs because of the stress on the wrists and shoulder joint complex.

MENTAL FOCUS FOR THE HOSE DRAG

Some examples of mental focus for the Hose Drag are:

- Visualize pulling the hose with smooth, rapid strokes. Pull the hose with a steady, rhythmic, hand-over-hand motion.
- While you are transitioning from the Step Mill, focus on lowering your heart rate. Relax your arms at your sides and use deep breathing.

Successful candidates have a mental focus and plan for each event. Only you can determine what mental focus will work for you (see Chapter 3—Module 4 on mental training).

Figure M6-2 Hose carry.

Figure M6-3 Hose pull.

SKILL DEVELOPMENTAL EXERCISES FOR THE HOSE DRAG

Refer to Appendix A for illustrations and performance tips for the following exercises **(Figures M6-4, M6-5)**:

Lower Body Strength	Balance and Grip Strength	Upper Body Strength
Squats	BOSU balance sequence	db rowing
Leg press	Dyna Disc balance	db curls
Lunges	Walking backwards on the Step Mill **(Figure M6-4)**	Partial dead lifts
One-legged squats	Wrist curls	Cable one-arm kneeling pulls **(Figure M6-5)**
Bench step-ups with weight (10 reps with each leg)		

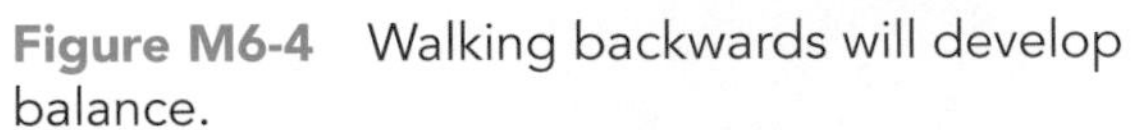

Figure M6-4 Walking backwards will develop balance.

Figure M6-5 Kneeling cable pulls.

PERFORMANCE PLANNING CHART

Event 2: Hose Drag

GOALS

- Using the SMART technique, list your learning and practice goals for the Hose Drag. What challenges are you facing and what specific actions do you need to do to meet these challenges? For each goal, list a completion date, objectives, and a feedback source.
- Evaluate each goal after completion. If necessary, set new goals to keep progressing.

LEARNING

Learning Points for the Hose Drag

- Walk to the drum and then to the box. Running will not save that much time and will raise your pulse.
- Although you will be on one knee, center yourself and keep your back in proper alignment. This will make your pulling motion more efficient.

What other learning points will lead to success in the Hose Drag?

____

____

Learning Goals

List how you will learn to do the Hose Drag:

Goal 1 ______________________ **Completion date:** __________

Objectives: ____

____

Feedback source: ____

Learning outcome: _____ **positive** _____ **neutral** _____ **negative**

Goal 2 ______________________ **Completion date:** __________

Objectives: ____

____

Feedback source: ____

Learning outcome: _____ **positive** _____ **neutral** _____ **negative**

PERFORMANCE PLANNING CHART

Event 2: Hose Drag—Cont'd

PRACTICE

What mental focus will you use while practicing for the Hose Drag?

____

____

Practice Goals

When and where will you practice? Detail how you will overlearn and use active practice in your training sessions for the Hose Drag.

Goal 1 ______________________ Completion date: __________

Objectives: ______________________________________

____

Feedback source: ______________________________________

Practice technique: ____ Active practice ____ Overlearning

Goal 2 ______________________ Completion date: __________

Objectives: ______________________________________

____

Feedback source: ______________________________________

Practice technique: ____ Active practice ____ Overlearning

Goal 3 ______________________ Completion date: __________

Objectives: ______________________________________

____

Feedback source: ______________________________________

Practice technique: ____ Active practice ____ Overlearning

MODULE 7
Equipment Carry

Figure M7-1 Many steps are involved in the Equipment Carry event.

KEY STATS

Average success time:	38 seconds
Average transition time:	18 seconds
Cardiovascular demand:	High
Muscles involved:	Upper back, biceps, deltoids, core body, all lower body muscles
Skills needed and abilities needed:	Balance, kinesthetic awareness of your body carrying two saws while in motion.
	Grip strength
	Anaerobic capacity
	Ability to use mental focusing techniques to perform all of the required tasks

TECHNIQUE AND STRATEGY

Mental focusing techniques are crucial for this event because of the many tasks involved. You will also need to develop good kinesthetic awareness of proper body position when carrying the two saws. Your practice sessions should be designed to overlearn the sequence of tasks and proper lifting techniques involved. Focus on centering yourself and avoiding any twisting, rotating, slumping, or rounding of posture. Keeping yourself centered will decrease excessive strain to your low back and reduce the chance of possible injury.

There are many steps involved in this event. Missing a step or performing a step out of sequence could lead to disqualification. The sequence of steps is as follows:

1. Remove the saws from the tool cabinet, one at a time, and place them on the ground.

2. Pick up both saws, one in each hand, and carry them while walking 75 ft around the drum and back to the starting point.
3. Place the saws on the ground in front of the cabinet.
4. Pick up each saw, one at a time, and replace it in the cabinet.

A key technique is to use your legs instead of your back when removing the saws from the cabinet. Remove them one at a time and place them on the ground. Turn around and face the drum before you pick up the saws. If you pick up the saws while facing the cabinet, the likelihood of twisting or rotating your torso when turning to face the drum will increase. This will place a lot of stress on your lower back and could lead to injury.

Your grip strength must be sufficient because you are not allowed to drop the saws at any time. You may set the saws down and readjust your grip if necessary.

The balance may be different for each saw so your practice sessions should include saws or objects with different balance points. You are not allowed to run during the event. Visualize walking with your body in a centered position, executing each step with flawless technique.

Figure M7-2 Turn and face the drum before picking up the equipment.

CPAT SPECIFIC EXERCISES FOR THE EQUIPMENT CARRY

Ideal Training Exercise

Place two saws in a cabinet 4 ft above ground level. Set up a drum 75 ft away. Remove the saws, one at a time, and place them on the ground. Turn and face the drum **(Figure M7-2)**, pick up the saws and carry them around the drum and back to the starting point. Place both saws on the ground and then replace them, one at a time, on the shelf. To overlearn this event, add weight to each saw once you have mastered the saw weight.

Alternate Training Exercise

Place two 15-lb dumbbells on a shelf 4 ft above ground level. Set up a drum or other marker (paint can, trash can) 75 ft away. Remove the weights, one at a time, and place them on the ground. Turn and face the drum, pick up the weights and carry them **(Figure M7-3)** around the drum and back

Figure M7-3 Keep yourself centered when carrying the dumbbells.

to the starting point. Place both weights on the ground and then replace them, one at a time, on the shelf.

Perform three repetitions then increase the weight of the dumbbells. Do not use dumbbells heavier than 30 lbs as this will place unnecessary stress on your joints.

Performance Point

Avoid twisting your torso when picking up the saws, the torque created by twisting your torso places excess stress on your lower back and could lead to injury. Remove the saws one at a time from the cabinet, then turn and face the drum before picking up the saws. This will help you maintain a centered position and minimize stress to your lower back.

Mental Focus for the Equipment Carry

Some candidates who fail the test are physically prepared but are lacking in mental focus. Examples of mental focus for the Equipment Carry are:

- Visualize yourself with strong trapezius or upper back muscles.
- Focus on keeping a centered position when carrying the saws.

Successful candidates have a mental focus and plan for each event. Only you can determine what mental focus will work for you (see Chapter 3—Module 4 on mental training).

Skill Developmental Exercises for the Equipment Carry

Refer to Appendix A for illustrations and performance tips for the following exercises **(Figures M7-4, M7-5)**:

Anaerobic Capacity	Balance, Core Body, and Hand Grip Strength	Upper and Lower Body Muscular Strength
Scott drills	BOSU plank holds BOSU balance sequence	Partial dead-lifts db press
Interval training	Varied hand position plank crawls	db bench press pushups
	Fit ball crunches Fit ball hyperextensions Fit ball Russian twist (see **Figures M7-4, M7-5**)	Leg press Lunges
	Crawling on stomach	Bench squats
	Contra lateral from kneeling position	
	Wrist curls	Close grip press
	Hand grip squeezes	

Figure M7-4 Russian twist—starting position.

Figure M7-5 Russian twist—finishing position.

PERFORMANCE PLANNING CHART

Event 3: Equipment Carry

GOALS

- Using the SMART technique, list your learning and practice goals for the Equipment Carry. What challenges are you facing and what specific actions do you need to do to meet these challenges? For each goal, list a completion date, objectives, and a feedback source.
- Evaluate each goal after completion. If necessary, set new goals to keep yourself progressing.

LEARNING

Learning Points for the Equipment Carry

- Keep your body centered. Do not slouch, twist, or rotate your torso when carrying the saws.
- Turn around and face the drum before you pick up the saws.
- Count the number of crawls and know at what count the ceiling lowers or wall narrows.

What other learning points will lead to success in the Search event?

____

____

Continued

PERFORMANCE PLANNING CHART

Event 3: Equipment Carry—Cont'd

Learning Goals

List how you will learn to do the Equipment Carry.

Goal 1 ______________________ Completion date: __________

Objectives: ______________________________

____

Feedback source: ______________________________

Learning outcome: _____ positive _____ neutral _____ negative

Goal 2 ______________________ Completion date: __________

Objectives: ______________________________

____

Feedback source: ______________________________

Learning outcome: _____ positive _____ neutral _____ negative

Goal 3 ______________________ Completion date: __________

Objectives: ______________________________

____

Feedback source: ______________________________

Learning outcome: _____ positive _____ neutral _____ negative

PERFORMANCE PLANNING CHART

Event 3: Equipment Carry—Cont'd

PRACTICE

- What mental focus will you use while practicing for the Equipment Carry?

____

____

Practice Goals

When and where will you practice? Detail how you will overlearn and use active practice in your training sessions for the Equipment Carry.

Goal 1 ______________________ **Completion date:** __________

Objectives: ______________________________

____

Feedback source: ______________________________

Practice technique: _____ **Active practice** _____ **Overlearning**

Goal 2 ______________________ **Completion date:** __________

Objectives: ______________________________

____

Feedback source: ______________________________

Practice technique: _____ **Active practice** _____ **Overlearning**

Goal 3 ______________________ **Completion date:** __________

Objectives: ______________________________

____

Feedback source: ______________________________

Practice technique: _____ **Active practice** _____ **Overlearning**

MODULE 8
Ladder Raise and Extension

Figure M8-1 Technique and power are used to raise the ladder.

KEY STATS

Average success time:	19 seconds
Transition time:	17 seconds
Cardiovascular demand:	Moderate
Muscles involved:	Hand grip strength, upper body muscles, and all lower body muscles
Skills and abilities needed:	Balance, core body strength, lower body power and endurance
	Ability to perform hand-over-hand method of raising ladder
	Ability to lock the rope when raising fly section of extension ladder

TECHNIQUE AND STRATEGY

This event requires the most hand-eye coordination of all the CPAT events.

When raising the ladder, use the hand-over-hand technique to walk it to an upright position against the wall **(Figure M8-1)**. Focus on using every rung during the raise.

Keep your body centered and apply the power of your lower body when raising the ladder. Visualize the force generated by your lower body being transferred into the action of raising the ladder. Use the weight and momentum of the ladder to your advantage by taking short steps. Tilt the ladder toward the wall when you have finished raising it.

Go immediately to the ladder extension. When extending the ladder, use the hand-over-hand method until it hits the stop. **Lock the rope (Figure M8-2)** by grabbing it with your palm facing away from you (pinkie on top), then turn the palm of your hand toward you **(Figure M8-3)**. This will prevent it from slipping through your

Figure M8-2 Locking the rope—starting position.

Figure M8-3 Locking the rope—finishing position.

hands. Keep the rope directly in front of you and do not raise it higher than your head.

Once the extension has hit the top, lower it by using the same hand-over-hand, locking technique. Keep your feet inside the boundary to avoid a warning.

> **Performance Point**
>
> When you complete the ladder raise, move immediately to the ladder extension. Use visualization in your training sessions to ensure that the transition is smooth. Overlearn the transition so that you do not waste any time starting the ladder extension. Practice exactly how and where you will grab the rope.

CPAT SPECIFIC EXERCISES FOR THE LADDER RAISE AND EXTENSION EVENT

Ideal Training Exercise

Ladder Raise—Use a 24-ft aluminum extension ladder. Anchor the ladder against a building to prevent it from sliding or falling when you are raising it.

Start slowly raising the ladder using the hand-over-hand method. Focus on hitting every rung. Increase your speed when you have overlearned the movement.

Ladder Extension—Use the same type of ladder that is used in the test. Use the hand-over-hand, locking the rope technique to raise the ladder to the stop. Using the same technique, lower the ladder in a controlling manner until it reaches the ground. Complete four repetitions, recording your time for each.

> **Performance Point**
>
> Mentally focusing on each rung of the ladder will help you develop the hand-over-hand technique of raising the ladder. Practice this skill slowly at first and increase your speed when you feel ready.

Alternate Training Exercise

Ladder Raise—Raise a 12-ft extension ladder to the top by using the hand-over-hand technique. Lower it back to the starting position on the ground.

Perform this exercise two times. Then proceed to the ladder extension.

Ladder Extension—Attach a rope to a weighted duffel bag. Place the rope over a tree branch or horizontal bar about 10 ft above the ground. With hand-over-hand movements raise the bag to the top and then slowly lower it to the ground. Be sure to use the rope locking technique. Start with a light weight and increase until you reach 40 lbs.

MENTAL FOCUS FOR THE LADDER RAISE AND EXTENSION

Some examples of mental focus for the Ladder Raise and Extension are:

- Focus on each rung as you raise the ladder. How many rungs must you touch before the ladder is upright?
- Develop a mantra for each task. Examples: "quick hands," "strong pull," etc.

SKILL DEVELOPMENTAL EXERCISES FOR THE LADDER RAISE AND EXTENSION

Refer to Appendix A for illustrations and performance tips for the following exercises **(Figures M8-4, M8-5)**:

Lower Body Power and Strength	Core Body Strength and Endurance; Balance	Upper Body Pushing and Pulling Strength and Endurance; Hand Grip Strength	Hand-over-Hand Raising Method; Locking the Rope
Squats	Fit Ball crunch sequence	db press	Core pole locking technique
Leg press	BOSU balance sequence	Shrugs	Elastic tube locking technique
Partial dead lifts	Walking lunges	Partial dead lifts	Cable hand-over-hand extension
Seated and standing calf raises	Floor plank	db curls	Elastic tube raising technique **(Figure M8-4)**
	BOSU plank	Wrist rollers	
		Tricep extension **(Figure M8-5)**	
		db incline press	

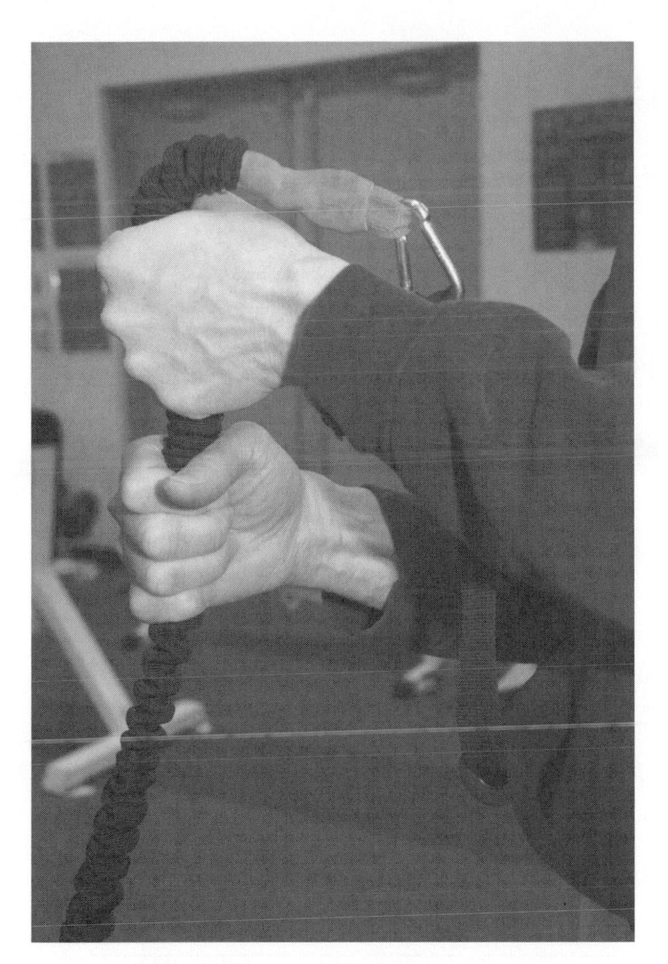

Figure M8-4 Elastic tube raise.

Figure M8-5 Tricep extension.

PERFORMANCE PLANNING CHART

Event 4: Ladder Raise and Extension

GOALS

- Using the SMART technique, list your learning and practice goals for the Ladder Raise and Extension. What challenges are you facing and what specific actions do you need to do to meet these challenges? For each goal, list a completion date, objectives, and a feedback source.
- Evaluate each goal after completion. Set new goals to keep progressing, if necessary.

LEARNING

Learning Points for the Ladder Raise and Extension Event

- Keep your body centered and apply the power of your lower body when raising the ladder.
- Get the ladder moving and keep it moving by taking short steps. Use the momentum of the ladder to your advantage.
- When extending the ladder, lock the rope by grabbing the rope with your palm facing away from you, then turning the palm of your hand toward you.

What other learning points will lead to success in the Ladder Raise and Extension event?

____

____

Learning Goals

List how you will learn to do the Ladder Raise and Extension:

Goal 1 ______________________ **Completion date:** __________

Objectives: ____

____

Feedback source: ____

Learning outcome: ____ **positive** ____ **neutral** ____ **negative**

Goal 2 ______________________ **Completion date:** __________

Objectives: ____

____

Feedback source: ____

Learning outcome: ____ **positive** ____ **neutral** ____ **negative**

PERFORMANCE PLANNING CHART

Event 4: Ladder Raise and Extension—Cont'd

PRACTICE

What mental focus will you use while practicing for the Ladder Raise and Extension?

____

____

Practice Goals

When and where will you practice? Detail how you will overlearn and use active practice in your training sessions for the Ladder Raise and Extension.

Goal 1 ______________________ Completion date: __________

Objectives: ______________________________________

____

Feedback source: ______________________________________

Practice technique: _____ Active practice _____ Overlearning

Goal 2 ______________________ Completion date: __________

Objectives: ______________________________________

____

Feedback source: ______________________________________

Practice technique: _____ Active practice _____ Overlearning

Goal 3 ______________________ Completion date: __________

Objectives: ______________________________________

____

Feedback source: ______________________________________

Practice technique: _____ Active practice _____ Overlearning

MODULE 9
Forcible Entry

Figure M9-1 Good instruction will help you perfect your swing.

KEY STATS

Average success time:	12 seconds
Transition time:	18 seconds
Cardiovascular demand:	High
Muscles involved:	Chest and arms, upper and lower back, deltoids, quadriceps, hamstrings, glutes, calves
Skills needed:	Balance; upper and lower body strength and endurance
	Ability to apply upper and lower body power when swinging the sledge hammer
	Grip strength

TECHNIQUE AND STRATEGY

This event requires explosive power and proper technique **(Figure M9-1)**. Each swing must drive the target inward because the average time of successful candidates is 12 seconds. The mental skills of focusing and visualization are crucial. Visualize yourself performing with perfect execution; seeing the force being transferred from your body to the target. Focus on the target and swing the hammer just like baseball players swing a bat by stepping forward with your lead foot. Be careful not to step in the toe box. When the alarm goes off, focus on dropping the hammer and not flinging or throwing it as this will result in a failure.

You can overlearn this event by increasing the weight of the sledge hammer to 12 lbs. Do not use a heavier weight because of the stress it will place on your shoulders.

> **Performance Point**
>
> Use the explosive power developed by two skill developmental exercises, the leg press and squat, to help you in this event. When performing these exercises, visualize yourself performing the Forcible Entry event with power and flawless technique. When you actually take the test, it will be the finalization of something you have already done.

CPAT SPECIFIC EXERCISES FOR THE FORCIBLE ENTRY

Ideal Training Exercise

Wrap a large amount of padding around a sturdy vertical pole or object. Draw a circle target on the padding 39 inches off the ground. Using a 10-lb sledge hammer, center yourself and focus on using the power generated from your legs and hips to swing the hammer so that the head strikes the target. Perform four sets of 10 repetitions. As your strength increases and you develop kinesthetic awareness of the swinging motion, try to increase the power generated.

Alternate Training Exercise

Perform cable swings and rope ball swings.

Cable swings—Using a cable crossover machine, set the cable at 39 inches above the ground. Using a "D handle," stand sideways, grasp the handle, and swing the cable using both hands **(Figure M9-2)**.

You should center yourself and "explode" into the swing. Go slow on the return of the cable to the starting position. Start with a light weight to test the motion, as every cable machine is different. Perform 10 repetitions, then rest and repeat. Do four sets of 10 repetitions, increasing the weight gradually. Do not use excessive weight on this exercise as you could easily injure the shoulder complex.

Rope ball swings—This exercise uses a medicine ball with a rope attached to it.

Grasp the rope and swing the ball against a brick or cement wall **(Figure M9-3)**. Use the full power of your body by centering and then "explode" into the motion. Control the ball on the way back, reset your stance and repeat. Perform four sets of 10 repetitions.

Figure M9-2 Cable swings.

MENTAL FOCUS FOR THE FORCIBLE ENTRY

Some candidates who fail the test are physically prepared but are lacking in mental focus. Examples of mental focus for the Forcible Entry are:

- Focus on generating power from your legs and "exploding" into the movement, transferring the power into the sledge hammer.

Figure M9-3 Medicine ball rope swing.

- Visualize yourself getting stronger with each repetition. With each stroke, try to swing the hammer through the measuring device.

Successful candidates have a mental focus and plan for each event. Only you can determine what mental focus will work for you (see Chapter 3—Module 4 on mental training).

SKILL DEVELOPMENTAL EXERCISES FOR THE FORCIBLE ENTRY

Refer to Appendix A for illustrations and performance tips for the following exercises:

Upper Body Strength and Endurance	Lower Body Strength and Endurance	Core Body and Balance; Grip Strength	Anaerobic Conditioning
db press	Squats	Fit ball abdominal sequence	Intervals
Pull down to front	Leg press	BOSU plank **(Figure M9-4)** floor plank	Scott drills
db curl	One-legged bench squats **(Figure M9-5)**	BOSU balance sequence	Sprints
Close grip press	Bench jumps	Wrist curls Wrist rollers	Lactate threshold intervals
Bench dips			
Chin-ups			
db rowing and partial dead lifts			

Figure M9-4 The plank develops strong abdominals.

Figure M9-5 One-legged bench squats.

PERFORMANCE PLANNING CHART

Event 5: Forcible Entry

GOALS

- Using the SMART technique, list your learning and practice goals for the Forcible Entry. What challenges are you facing and what specific actions do you need to do to meet these challenges? For each goal, list a completion date, objectives, and a feedback source.
- Evaluate each goal after completion. If necessary, set new goals.

LEARNING

Learning Points for the Forcible Entry

- Center yourself then explode into each repetition, using your entire body strength to swing the sledge hammer.
- Try to swing the sledge hammer through the measuring device.

What other learning points will lead to success in the Forcible Entry?

____

____

Learning Goals

List how you will learn to do the Forcible Entry:

Goal 1 ______________________ **Completion date:** __________

Objectives: ____

____

Feedback source: ____

Learning outcome: _____ **positive** _____ **neutral** _____ **negative**

Goal 2 ______________________ **Completion date:** __________

Objectives: ____

____

Feedback source: ____

Learning outcome: _____ **positive** _____ **neutral** _____ **negative**

Continued

PERFORMANCE PLANNING CHART

Event 5: Forcible Entry—Cont'd

PRACTICE

What mental focus will you use while practicing for the Forcible Entry?

____

____

Practice Goals

When and where will you practice? Detail how you will overlearn and use active practice in your training sessions.

Goal 1 ______________________ Completion date: __________

Objectives: ____

____

Feedback source: ____

Practice technique: _____ Active practice _____ Overlearning

Goal 2 ______________________ Completion date: __________

Objectives: ____

____

Feedback source: ____

Practice technique: _____ Active practice _____ Overlearning

Goal 3 ______________________ Completion date: __________

Objectives: ____

____

Feedback source: ____

Practice technique: _____ Active practice _____ Overlearning

MODULE 10
Search

Figure M10-1 Visualization will help you get through the tunnel.

KEY STATS

Average success time:	34 seconds
Average transition time:	21 seconds
Cardiovascular demand:	Moderate
Muscles involved:	Triceps, deltoids, chest, core body, lower body
Skills and abilities needed:	Good kinesthetic awareness of your body in a confined space Balance in a crawling position, core body strength Ability to use mental focusing techniques to overcome a lack of visual feedback

TECHNIQUE AND STRATEGY

Mental focusing techniques are paramount for the Search event because you can not see where you are going or what is coming up **(Figure M10-1)**. You will need to develop good kinesthetic awareness of how your body feels when crawling without visual feedback.

It is best to take a trial run through so that you can count the number of crawls and at what point the tunnel changes. This will allow you to find out where the obstacles are, when the walls begin to narrow, and at what point the ceiling starts to drop down so that you will have to crawl on your stomach.

You can also use the feedback generated from your helmet touching the top of the maze ceiling to judge how low you should crawl. When you feel the ceiling sloping down, you will have to crawl on your belly. Use the feedback from your hands touching the sidewalls to gauge the width of the tunnel **(Figure M10-2)**. Use one hand as "lead" to guide you through. There are no abrupt edges that you will run into and all the walls are tapered. At two points the tunnel will turn to your right.

It is important to keep moving so you do not lose precious seconds. As you crawl, balance the weight between both knees and hands. Maintain a smooth crawling motion and keep moving.

If you have claustrophobia, keep your mind on crawling and remember the exit is only a few seconds away.

CPAT SPECIFIC EXERCISES FOR THE SEARCH

Ideal Training Exercise

The best approach is to take a trial run on the actual tunnel to familiarize yourself with the layout and obstacles. Counting the crawls will tell you when changes are coming and help you prepare a mental focus for each one.

Alternate Training Exercise

Set up a U-shaped 64-ft tunnel with exercise mats. The ceiling of the tunnel should be 3 ft high and the walls 4 ft wide. Place a number of obstacles at various locations throughout the tunnel. They should force you to maneuver over, around, and under them. At two locations narrow the width of the tunnel, forcing yourself to go through a narrowed space. Lower the ceiling at several points to force yourself to crawl on your stomach for about 10 ft each.

Figure M10-2 Let feedback from your hand guide you through the tunnel.

Figure M10-3 Crawling blindfolded will develop kinesthetic awareness.

Track your time and when you have mastered the average success time (34 seconds), start practicing with a weight vest while blindfolded **(Figure M10-3)**. Also, begin to develop a mental focus for each obstacle as well as kinesthetic awareness for the crawling motion. If you do not have access to a weight vest, use a knapsack or backpack. Gradually increase the weight in the backpack until it reaches 50 lbs. One disadvantage to using a backpack instead of a vest is that it will scrape against the ceiling when you are crawling on your stomach.

Performance Point

Get in and keep moving. Your hands and helmet will let you know the height and width of the tunnel as you go through. Mental focusing is crucial for this event because of the lack of visual feedback.

MENTAL FOCUS FOR THE SEARCH

Some candidates who fail the test are physically prepared but are lacking in mental focus. Examples of mental focus for the Search are:

- When you encounter changes in the tunnel, use visualization techniques to move yourself trough the change.
- Visualize yourself crawling quickly all the way through.
- Try not to fight or beat the tunnel—experience it.
- When doing a CPAT trial run, count the number of "crawls." Break the total number down into sets and reps.

Successful candidates have a mental focus and plan for each event. Only you can determine what mental focus will work for you (see Chapter 3—Module 4 on mental training).

SKILL DEVELOPMENTAL EXERCISES FOR THE SEARCH EVENT

Refer to Appendix A for illustrations and performance tips for the following exercises **(Figure M10-4)**:

Mental Training Techniques	Balance in a Crawling Position	Muscular Strength and Endurance in a Crawling Position
Visualization of the tunnel while crawling blindfolded	BOSU plank holds	db press
Focusing point for each change in the tunnel	Varied hand position plank crawls	db bench press and push-ups
	Forward and reverse crawls	Fit Ball crunches, twists
	Crawling on stomach	Fit Ball hyperextensions
	Contra lateral from kneeling position	Lunges
	Resisted crawl **(Figure M10-4)**	Close grip press
		Leg press
		One-legged squats

Figure M10-4 Adding resistance will develop your crawling ability.

PERFORMANCE PLANNING CHART

Event 6: Search

GOALS

- Using the SMART technique, list your learning and practice goals for the Search event. What challenges are you facing and what specific actions do you need to do to meet these challenges? For each goal, list a completion date, objectives, and a feedback source.
- Evaluate each goal after completion. If necessary, set new goals to keep yourself progressing.

LEARNING

Learning Points for the Search Event

- Use your hands and helmet to gauge changes in the height and width of the tunnel.
- Balance your body weight between your shoulders and knees when crawling. Keep pressure off your wrists and shoulders.
- Count the number of crawls and know at what count the ceiling lowers or wall narrows.

What other learning points will lead to success in the Search event?

____

Learning Goals

List how you will learn to do the Search event:

Goal 1 ____________________ **Completion date:** __________

Objectives: ____________________

____

Feedback source: ____________________

Learning outcome: ____ **positive** ____ **neutral** ____ **negative**

Goal 2 ____________________ **Completion date:** __________

Objectives: ____________________

____

Feedback source: ____________________

Learning outcome: ____ **positive** ____ **neutral** ____ **negative**

Goal 3 ____________________ **Completion date:** __________

Objectives: ____________________

____

Feedback source: ____________________

Learning outcome: ____ **positive** ____ **neutral** ____ **negative**

PERFORMANCE PLANNING CHART

Event 6: Search—Cont'd

PRACTICE

What mental focus will you use while practicing for the Search event?

____

____

Practice Goals

When and where will you practice? Detail how you will overlearn and use active practice in your training sessions for the Search event.

Goal 1 ______________________ **Completion date:** __________

Objectives: ____

____

Feedback source: ____

Practice technique: _____ **Active practice** _____ **Overlearning**

Goal 2 ______________________ **Completion date:** __________

Objectives: ____

____

Feedback source: ____

Practice technique: _____ **Active practice** _____ **Overlearning**

Goal 3 ______________________ **Completion date:** __________

Objectives: ____

____

Feedback source: ____

Practice technique: _____ **Active practice** _____ **Overlearning**

MODULE 11
Rescue

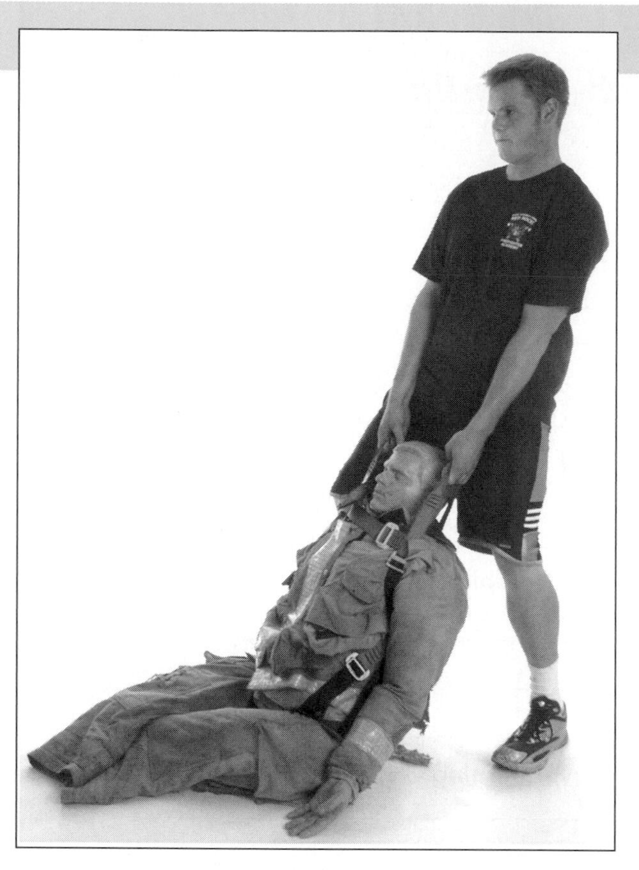

Figure M11-1 Use your back and leg muscles to pull the mannequin.

KEY STATS

Average success time:	31 seconds
Transition time:	21 seconds
Cardiovascular demand:	High
Muscles involved:	Hand grip strength, upper and lower back, all lower body muscles
Skills and abilities needed:	Balance; core body strength, lower body power and endurance
	Ability to keep the dummy in motion by using its own momentum

TECHNIQUE AND STRATEGY

This is one event that a lot of candidates have problems with. One candidate's times for the first six events were in the success range. However, the dummy drag took him 1:08, which is more than double the average success time for this event. The candidate ran out of time on the last event.

This event will challenge all of the major muscles in your body.

Successful candidates use the strongest muscles in their bodies to pull the dummy. They use their arms as "hooks" and the legs and back to keep the dummy moving by taking short, quick steps **(Figure M11-1)**. If you stop, it will require energy and strength to get the dummy moving again. Your strongest position is when you are centered: your chest is out, the back maintains a normal curve, and the knees are slightly bent **(Figure M11-2)**. This

Figure M11-2 Centering the body will allow you to maximize your strength.

Figure M11-3 Rounding the back will diminish your strength and increase your risk of injury.

position allows you to use the full strength of the legs and back—the strongest muscles in the body to pull the dummy. Avoid rounding your back. Not only will this make you weaker but will place a lot of stress on your lower back **(Figure M11-3)**.

> **Performance Point**
>
> Keep the dummy moving by thinking of the barrel as a gigantic magnet pulling you toward it. When you reach the barrel, make sure you pull most of the dummy past the barrel before making the turn. If you turn too soon, the dummy will drag against the barrel. Once you have turned, think of the finish line as a gigantic magnet pulling you toward it.

CPAT SPECIFIC EXERCISES FOR THE RESCUE EVENT

Ideal Training Exercise

Train with a 165-lb dummy. Start by pulling the dummy 35 ft and then circle around a pre-positioned drum. Make sure you do not grab or rest on the drum as this will result in disqualification. Pull the dummy 35 ft back to the starting point. If you have to stop and rest, build up your distance until you complete the entire test. Once you have completed the test without stopping, wear a weight vest and follow the same procedure until you complete the test without stopping. To overlearn this event, drag the dummy 50 ft before returning to the starting position. Compare your time to the average success time. Your goal should be to not lose a great deal of time on this event.

Alternate Training Exercise

Use a duffel bag with a handle and load the bag with 50 lbs of rocks, sand, or other material. Grab the handle, center yourself, and lock your arms into position. Start dragging the victim using your legs and back to do most of the work. Taking short steps, drag the bag 35 ft, turn around, and drag the bag back to the starting point. Keep increasing the weight in the bag with each attempt until you reach 165 lbs. Once you have mastered this, drag the bag 50 ft in each direction. This will produce an overlearning effect.

Performance Point

Your strongest position is when you are centered, with both hands grasping the dummy's handle. Pull the dummy with your back to the barrel. Some candidates pull the dummy with one hand, standing sideways or facing the drum. This technique requires an individual with above-average overall body and grip strength. However, it does not use the full strength of the individual and places a lot of strain on the lower back to stay in position.

MENTAL FOCUS FOR THE RESCUE

Examples of mental focus for the Rescue event are:

- Count the number of steps it takes you to complete the event. Break the total number of steps down into sets of 10 or 8. If your total steps equal 50, for example, it is easier to complete five sets of 10 steps than complete one set of 50 steps.
- Picture your legs as pistons for a steam locomotive train. See yourself going strong all the way through, with power and strength.
- Focus on the momentum of the dummy to keep it moving.

Successful candidates have a mental focus and plan for each event. Only you can determine what mental focus will work for you (see Chapter 3—Module 4 on mental training).

SKILL DEVELOPMENTAL EXERCISES FOR THE RESCUE

Refer to Appendix A for illustrations and performance tips for the following exercises **(Figures M11-4, M11-5)**:

Lower Body Power and Strength	Core Body Strength and Endurance; Balance	Upper Body Strength and Endurance	Cardiovascular Conditioning
Squats	Fit Ball crunch sequence	db rowing	Intervals and sprints
Leg press	BOSU balance sequence	Shrugs	Scott drills
Partial dead lifts	Walking lunges	Partial dead lifts	Up and down LT intervals
	Floor plank **(Figure M11-4)**	db curls	
	BOSU plank **(Figure M11-5)**	Wrist rollers	

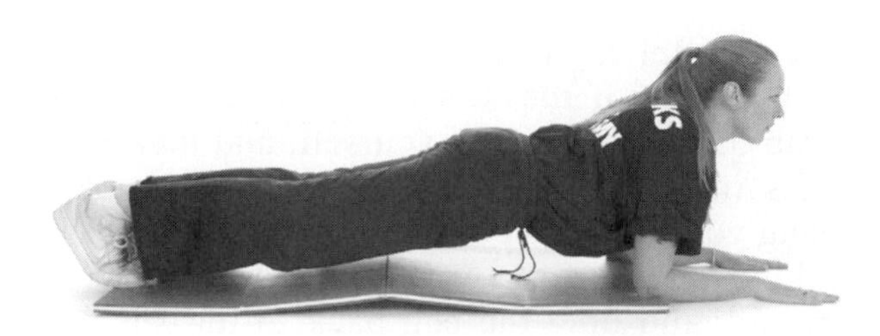

Figure M11-4 Floor plank.

Figure M11-5 BOSU plank.

PERFORMANCE PLANNING CHART

Event 7: Rescue

GOALS

- Using the SMART technique, list your learning and practice goals for the Rescue event. What challenges are you facing and what specific actions do you need to do to meet these challenges? For each goal, list a completion date, objectives, and a feedback source.
- Evaluate each goal after completion. Set new goals to keep progressing, if necessary.

LEARNING

Learning Points for the Rescue Event

- Maintain a "centering position." Keep your body in proper alignment: chest out, shoulders back, knees bent.
- Pull the dummy with your entire body. Think of your arms as hooks only, use the strength in your back and legs to do the work.
- Get the dummy moving and keep it moving by taking short steps. Use the momentum of the dummy to your advantage.
- Prepare yourself mentally on the transition to this event. Locate where the handles are and visualize yourself keeping the dummy in motion, pulling strong all the way through.

What other learning points will lead to success in the Rescue event?

____

Learning Goals

List how you will learn to do the Rescue event:

Goal 1 ______________________ **Completion date:** __________

Objectives: ______________________________

____

Feedback source: ______________________________

Learning outcome: _____ **positive** _____ **neutral** _____ **negative**

Goal 2 ______________________ **Completion date:** __________

Objectives: ______________________________

____

Feedback source: ______________________________

Learning outcome: _____ **positive** _____ **neutral** _____ **negative**

Continued

PERFORMANCE PLANNING CHART

Event 7: Rescue—Cont'd

PRACTICE

What mental focus will you use while practicing for the Rescue?

____

____

Practice Goals

When and where will you practice? Detail how you will overlearn and use active practice in your training sessions for the Rescue event.

Goal 1 ______________________ Completion date: __________

Objectives: ____

____

Feedback source: ____

Practice technique: _____ Active practice _____ Overlearning

Goal 2 ______________________ Completion date: __________

Objectives: ____

____

Feedback source: ____

Practice technique: _____ Active practice _____ Overlearning

Goal 3 ______________________ Completion date: __________

Objectives: ____

____

Feedback source: ____

Practice technique: _____ Active practice _____ Overlearning

MODULE 12

Ceiling Breach and Pull

Figure M12-1 Use the strength from your entire body to move the pike pole.

KEY STATS

Average success time:	46 seconds
Cardiovascular demand:	High
Muscles involved:	Biceps, triceps, deltoids, upper back, entire lower body, core body strength
Skills and abilities needed:	Balance; upper and lower body strength and endurance
	Ability to apply upper and lower body power to push and pull the ceiling apparatus
	Grip strength

TECHNIQUE AND STRATEGY

Many candidates consider the Ceiling Breach and Pull to be the hardest event because they are fatigued from performing the previous events.

On the transition from the Rescue event, visualize yourself grasping the pike pole and performing the event smoothly. Use deep breathing to prepare yourself physically. Because you have overlearned this event, you will not waste precious seconds pausing to think how to perform the event. When you get to the event start right in.

You need to do four sets of three pushes and five pulls.

Keep the pike pole in motion. This will require less energy as each time you stop it takes more effort to get the pole moving again. Use the momentum of the pole and the strength from your entire body to raise and lower the pike pole **(Figure M12-1)**.

Most unsuccessful candidates try to complete this event by using only their arms. Using your arms only will not generate enough power to push the hinged door or pull the ceiling device, and this will cost precious seconds.

Think of your arms as attachments. The force to pull and breach the ceiling must come from a centered position. This uses the strength from your entire body. You can tell if you are applying enough power if you hear the ceiling hatches making noise as you push and pull.

Performance Point

Grab the pike pole as high as you can and use your body and vest weight when pulling the pole down. Lock your arms and use the force generated from your lower body to pull the pole down. This technique will help save your strength.

CPAT SPECIFIC EXERCISES FOR THE CEILING BREACH AND PULL

Ideal Training Exercise

Ceiling pull—Perform one-arm pull downs **(Figure M12-2)**. Using an adjustable cable machine, place the cable at the high setting and set the weight at 10 to 15 lbs. Kneel down so you will have plenty of room to extend your arm. Start with your arm in an almost extended position and pull the handle down to your mid-section. Pull down with power then return to the top position with control. Start with a light weight, gradually increasing until you reach 80 lbs (the weight of the ceiling device). Using more weight will place too much stress on your joints and wrists. Perform 10 repetitions, then repeat with the opposite hand.

Ceiling breach—Place the cable at the low setting, start with a light weight, grasp the handle, and perform cable upright rows **(Figure M12-3)**. Stand holding the cable handle at waist level, quickly push the handle straight upward in front of your body, being careful not to hyperextend your elbow. Lower the weight gradually. Perform 10 repetitions, then repeat using the opposite hand. It is important to center yourself and use power to push the handle up and to control the weight on the way down. Use a count of up 1-2 and down 1-2-3-4. Perform 10 repetitions, then repeat with the opposite hand.

Figure M12-2 One-arm pull down.

Alternate Training Exercise

Ceiling pull—Attach a rope to a dumbbell, weight plate, or knapsack. Place the rope over a sturdy horizontal bar about 10 ft off the ground. Grab the rope and position your hands about a foot apart with the lower hand about chin level. Center yourself and explode into the movement, using the power generated from your core and legs to raise the weight; lower the weight with control. Perform three sets of 10 repetitions with a light resistance, increasing the weight until you reach 80 lbs.

Ceiling breach—Attach a rope to a dumbbell or weighted knapsack. Grasp the rope with one hand about waist level and the other about 1 ft higher. Center yourself and explode upward, using your lower body power to drive the rope up. Lift upward as high as you can, being careful not to hyperextend your elbows; lower the weight with control. Start with a light weight and increase until you reach 60 lbs (the weight of the

Figure M12-3 Upright rows.

hinged door). Perform 10 repetitions, reverse hand positions then repeat. Complete three of these cycles.

MENTAL FOCUS FOR THE CEILING BREACH AND PULL

Some candidates who fail the test are physically prepared but are lacking in mental focus. Examples of mental focus for the Ceiling Breach and Pull are:

- Focus on generating power from your legs as much as possible. Picture your legs as pistons when pushing and pulling the pike pole.
- Visualize yourself getting stronger with each repetition. With each stroke try to push the pike pole through the top of the hinged door.

Successful candidates have a mental focus and plan for each event. Only you can determine what mental focus will work for you (see Chapter 3—Module 4 on mental training).

SKILL DEVELOPMENTAL EXERCISES FOR THE CEILING BREACH AND PULL

Refer to Appendix A for illustrations and performance tips for the following exercises **(Figures M12-4, M12-5)**:

Upper Body Strength and Endurance	Lower Body Strength and Endurance	Core Body Strength	Anaerobic Conditioning
press db	Squats	Fit ball abdominal sequence	Intervals
Pull down to front	Leg press	BOSU plank Floor plank	Scott drills
db curl	One-legged bench squats		Sprints
Tricep extension	Bench jumps **(Figures M12-4, M12-5)**		Lactate threshold intervals
Bench dips			
Chin-ups			
db rowing			

Figure M12-4 Bench jumps—starting position.

Figure M12-5 Bench jumps—ending position.

PERFORMANCE PLANNING CHART

Event 8: Ceiling Breach and Pull

GOALS

- Using the SMART technique, list your learning and practice goals for the Ceiling Breach and Pull. What challenges are you facing and what specific actions do you need to do to meet these challenges? For each goal, list a completion date, objectives, and a feedback source.
- Evaluate each goal after completion. If necessary, set new goals.

LEARNING

Learning Points for the Ceiling Breach and Pull

- Center yourself, then explode into each repetition, using your entire body strength to move the pike pole upward.
- Keep the pike pole in motion. Use the momentum of the apparatus to your advantage.

What other learning points will lead to success in the Ceiling Breach and Pull?

____

____

Learning Goals

List how you will learn to do the Ceiling Breach and Pull:

Goal 1 ______________________ **Completion date:** __________

Objectives: ____

____

Feedback source: ____

Learning outcome: _____ **positive** _____ **neutral** _____ **negative**

Goal 2 ______________________ **Completion date:** __________

Objectives: ____

____

Feedback source: ____

Learning outcome: _____ **positive** _____ **neutral** _____ **negative**

Continued

PERFORMANCE PLANNING CHART

Event 8: Ceiling Breach and Pull—Cont'd

PRACTICE

What mental focus will you use while practicing for the Ceiling Breach and Pull?

____

____

Practice Goals

When and where will you practice? Detail how you will overlearn and use active practice in your training sessions.

Goal 1 ______________________ **Completion date:** __________

Objectives: ______________________________

____

Feedback source: ______________________________

Practice technique: ____ **Active practice** ____ **Overlearning**

Goal 2 ______________________ **Completion date:** __________

Objectives: ______________________________

____

Feedback source: ______________________________

Practice technique: ____ **Active practice** ____ **Overlearning**

Goal 3 ______________________ **Completion date:** __________

Objectives: ______________________________

____

Feedback source: ______________________________

Practice technique: ____ **Active practice** ____ **Overlearning**

Case Study: CPAT Trial Run

Firefighter candidates Angie and Bill took trial runs on the CPAT course.

Case 1

Angie's times for each event and transitions were:

Event time/Transition time (seconds)	
1. Step Mill	180/40
2. Hose Drag	37/26
3. Equipment Carry	50/26
4. Ladder Raise and Extension	30/25
5. Forcible Entry	72/25
6. Search	116/28
7. Rescue	146/38
8. Ceiling Breach and Pull	132/0
Total:	763/208
Total time in minutes:	16:11

She barely made it through the Step Mill and had to stop at all of the transitions to catch her breath. On the Search event, she stopped every time she encountered an obstacle. She experienced extreme fatigue on the last two events. On the Rescue event she stopped six times and on the Ceiling Breach and Pull she paused four times to catch her breath.

Case 2

Bill's times for each event and transitions were:

Event time/Transition time (seconds)

1. Stair Climb 180/50
2. Hose Drag and pull failure (failed to go outside of the drum)

Bill was exhausted after completing the Stair Climb event. He did not feel comfortable on the Step Mill and received two warnings for grabbing the handrail. He stopped to catch his breath twice on the transition to event 2.

On the Hose Drag and pull he failed to go outside of the drum when running to the marked box.

Case Study Questions

1. What goals would help Angie and Bill improve their CPAT performance?
2. What is the **major** reason Bill failed on his practice CPAT run?
3. How will mental focusing help Angie improve her CPAT performance?

CHAPTER SUMMARY

Candidates must develop a variety of physical abilities and mental focusing skills to pass the CPAT. All the events require candidates to possess excellent cardiovascular conditioning, with the most demanding event being the Step Mill. Upper and lower body muscular strength, balance, and core body strength are also required for success.

In addition, there are specific skill requirements that the candidate must demonstrate to be successful in each event.

Centering allows the candidate to transfer the explosive power generated from the lower body and core into the execution of the events. Prime examples of centering are pulling the dummy in the Rescue event, swinging the sledge hammer in the Forcible Entry event, and the raising and pulling of the hinged door in the Ceiling Breach and Pull event. Using the centering technique when performing the Ceiling Breach and Pull will also help to overcome fatigue.

Kinesthetic awareness is the feedback from movements of the body. This skill is required in the Search event because of the lack of visual feedback and the constraints involved in crawling in a confined space. Kinesthetic awareness of the stair climbing motion will allow candidates to maintain balance and develop rhythm in the Step Mill event.

A high level of mental focusing abilities will allow candidates to successfully complete the many tasks involved in the Equipment Carry and Search events.

Grip strength and hand coordination are key requirements when performing several events.

Examples are the pulling motion necessary for the Hose Drag and pull event, the hand-over-hand method of raising the ladder and the ability to "lock" the rope in place when raising and lowering the ladder in the Ladder Raise and Extension.

Using the momentum of the objects in motion when performing the events will enhance a candidate's abilities in the CPAT. Starting and stopping increases energy demands and will result in fatigue.

CHECK YOUR LEARNING

1. What is the best position for your hands during the Step Mill event?
   - **a.** Holding on to the side rails.
   - **b.** Grasping the vest in front of your body.
   - **c.** Hanging down at your sides.
   - **d.** b and c
   - **e.** None of the above.

2. The most effective way to swing the sledge hammer during the Forcible Entry event is
   - **a.** To stand as close as possible to the target.
   - **b.** To stand as far as possible from the target.
   - **c.** To use the force generated from your whole body by centering.
   - **d.** To use the strength from your arms to generate force.

3. Locking the rope during the Ladder Raise and Extension refers to
   - **a.** Wrapping the rope three times around your hand before pulling.
   - **b.** Turning the palm upward with each pulling stroke.
   - **c.** Attaching the rope to your belt buckle.
   - **d.** Pulling the rope with both hands.

4. Which of the following will result in a failure during the Hose Drag event?
   - **a.** Running with the hose during the 100-ft transition to the marked box.
   - **b.** Pulling the hose with one knee in contact with the ground.
   - **c.** Pulling the hose with two hands.
   - **d.** Pulling the hose with your knee positioned outside the marked box.

5. Proper procedure for the Equipment Carry event is to
   - **a.** Upon return to the cabinet, place both saws on the ground, pick up one saw at a time and replace the saw back in the cabinet.
   - **b.** Carry both saws to the drum, place one of them on the ground, then return back to the cabinet and replace the other saw.
   - **c.** Remove both saws from the cabinet and start immediately toward the drum.
   - **d.** Drop either of the saws to the ground when returning from the drum.

6. The most effective way to raise and lower the pike pole during the Ceiling Breach and Pull event is to
   - **a.** Use the force generated from your arms for both the upward push and downward pull strokes.
   - **b.** Stop and rest if you feel fatigued, then resume raising and pulling the ceiling door.
   - **c.** Raise the door by centering yourself and using the force generated from your entire body; use your body weight and your arms only as hooks to pull down the door.
   - **d.** Move your feet outside the marked boundary box to gain more leverage.

7. Maximizing your power for the Rescue event involves
   - **a.** Pulling the dummy with one hand while moving sideways.
   - **b.** Locking your body into a centered position while pulling the dummy with your back to the drum.
   - **c.** Rounding your back while pulling the dummy with your back to the drum.
   - **d.** Pulling the dummy with one hand while facing the drum.

8. Kinesthetic awareness for the Search event refers to
- **a.** Keeping your knees in constant motion.
- **b.** Memorizing the location of the obstacles in the tunnel.
- **c.** Using the feedback generated from your hands and body in motion to guide your movement.
- **d.** Calling out to the proctor for directions to the tunnel exit.

9. Which of the following constitutes a failure for the Step Mill event?
- **a.** Dismounting the Step Mill once during the warm-up period.
- **b.** Touching the wall momentarily for balance during the test.
- **c.** Dismounting the Step Mill during the test.
- **d.** Grasping the hand rail once during the test.

10. Which of the following actions will result in a disqualification or failure?
- **a.** Running during the transitions between events.
- **b.** Walking during the transitions between events.
- **c.** Dropping to one knee to pull the hose in the Hose Drag event.
- **d.** Stopping to catch your breath during the transitions between events.

THE MASTER TRAINING SCHEDULE

LEARNING OBJECTIVES

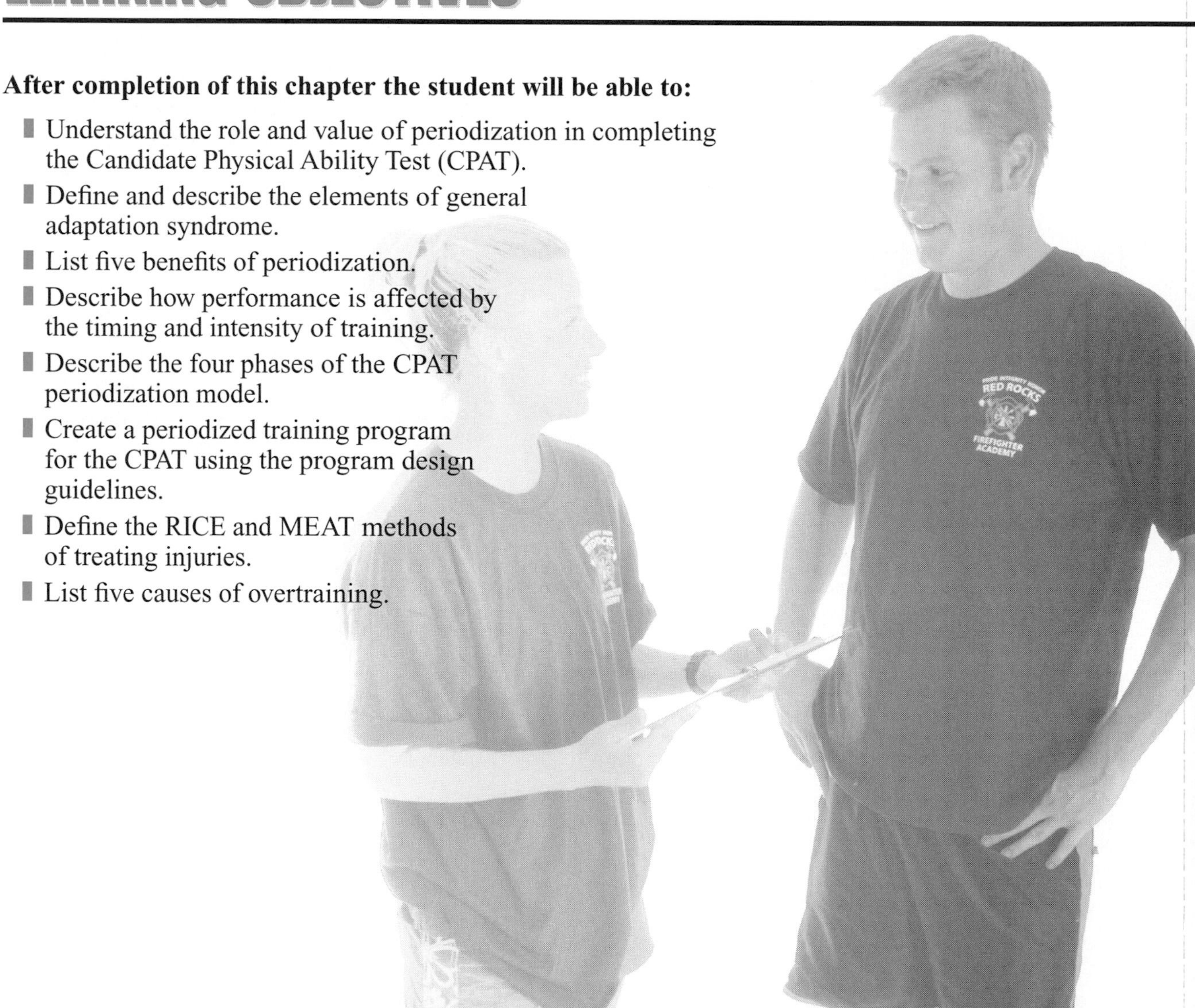

After completion of this chapter the student will be able to:

- Understand the role and value of periodization in completing the Candidate Physical Ability Test (CPAT).
- Define and describe the elements of general adaptation syndrome.
- List five benefits of periodization.
- Describe how performance is affected by the timing and intensity of training.
- Describe the four phases of the CPAT periodization model.
- Create a periodized training program for the CPAT using the program design guidelines.
- Define the RICE and MEAT methods of treating injuries.
- List five causes of overtraining.

INTRODUCTION

Preparing for the Candidate Physical Ability Test (CPAT) is much like training for a triathlon **(Figure 5-1)**. In fact, the similarities are striking. A triathlon consists of three events, with transition points between each event. The CPAT consists of eight events, with transition points between each event. Both disciplines require athletes to condition themselves in high- and low-intensity cardiovascular training, mental training, and specific training for each event. Both disciplines require training the transition points between events.

How do triathletes train to be in peak condition for all three events when it comes time for the competition? The challenge is to design a training program which addresses all these needs. To accomplish this most athletes use planned training strategies known collectively as **periodization**.

A periodized training plan will help you train concurrently for each of the eight CPAT events while integrating strength and other training modalities. It will help you plan training sessions in a progressive manner so that you will "peak" at the time of testing. This chapter will explain the principles of periodization and why they are crucial to your success on the CPAT. Training schedules are outlined in Appendix B. Included are 16, 12, 8, and 4 week detailed programs that will enable you to be at your best for the actual test.

Performance Point

Planning your training is essential to your success on the CPAT (see **Figure 5-1**). As the saying goes, "Failing to plan is planning to fail." Without a plan, you will have no idea where you stand on the road to success. Plans of 16, 12, 8, and 4 weeks are presented here. If you choose to develop your own plan, follow the principles listed under designing your own program.

Figure 5-1 Proper planning is essential to your success in the CPAT.

HISTORY OF PERIODIZATION

Periodization is a training method that seeks to "peak" an athlete's performance with a competitive event. This is accomplished by changing training intensity and strategically placing maintenance and recovery phases to enhance performance.

In the early 1900s, Canadian scientist Hans Selye developed a theory called the **general adaptation syndrome** which described an organism's response to stress. He defined a positive adaptation to stress, **eustress**, as being the result of correctly timed alternation between stress and regeneration. If there is too great a stimulus and/or too little regeneration, a negative adaptation, or **distress**, results.

Following a positive adaptation to the stress, the organism is capable of doing more work. This enhanced capability is referred to as **supercompensation**.[1,2] From this work, scientists in the former Eastern Bloc countries of the Soviet Union and East Germany evolved a theory of training called "periodization." It was not until the latter half of the 20th century that the idea of periodizing an athlete's training program became commonly accepted in the United States.[3]

Benefits of periodization include[4]:

- Produces better results than nonperiodized programs.
- Decreases the potential for overuse injuries.
- Keeps you fresh and progressing toward your goal.
- Enhances your compliance.
- Prevents training plateaus.
- Provides the rest and recovery that the mind and the body need to optimize training adaptations.

THE CPAT PERIODIZATION MODEL

The CPAT periodization model is divided into four phases of training **(Table 5-1)**. Each phase of training has its own objectives designed to progressively increase your fitness levels and CPAT skills.

Table 5-1. CPAT Periodization Model

Phase	Training Emphasis	Training Intensity
1	Base training	Low
2	Strength and endurance	Moderate
3	Strength, power and endurance	High
4	Peak power	Very high/low

Phase 1—Base Training

Objectives:

- Determine key baseline fitness measures.
- Set program goals.
- Learn the skills and techniques necessary to complete the CPAT stations.
- Learn the skills and techniques of resistance and cardiovascular training.
- Practice CPAT–specific skills.
- Develop a strong cardiovascular and strength base.

This is a learning phase. New skills and patterns of movement will be introduced to help you develop a strong training base **(Figure 5-2)**. Success in this phase results largely from neuromuscular changes and this will set the groundwork for the demands of subsequent phases. A gradual progression is important. If you begin training at a high level of intensity without laying down an adequate base, you will be susceptible to increased illness or injury. The amount of progress you can make later depends on how solid your training base is.

Phase 1 Training Guidelines

a. Complete all the biometric tests. This will provide you with baseline data on your fitness levels. Your starting points for all exercises will be determined by these results.

b. Fill in the performance planning chart from the fitness training module. Set your fitness goals using the information from the biometric testing sessions.

c. Plan out your resistance and cardiovascular training sessions. Strength training exercises should contain two sets of 12 repetitions. Performing this high repetition phase first will accelerate your strength potential in the later phases.

d. Cardiovascular intensity during this phase will equal zones 1 and 2.

e. Use the performance planning charts from the event modules to detail how you will learn the skills involved in each event. CPAT–specific intervals will last 60 seconds during this phase. Total high-intensity training time will equal about 15 minutes.

Figure 5-2 Good instruction will aid your learning in phase 1.

f. As you learn the skills, start to develop visual images of yourself executing the exercises and CPAT events with perfect form. Start to develop focus points and mantras for each CPAT event.
g. Functional skill training will involve exercises for strengthening the key abdominal and lower back areas. Balance training will also be addressed.
h. The last week of this phase is performed at a lower intensity of effort. It allows for physical recovery and enhancement of the adaptation process.

Phase 2—Strength and Endurance

Objectives:

- Build on the foundation developed in phase 1.
- Increase strength levels and endurance fibers.
- Develop smooth transitions between CPAT events.
- Increase mental focusing techniques.
- Evaluate your phase 1 goals.

In phase 2, you will build on the foundation developed in phase 1. Training intensity will increase, which will enable you to deal with higher blood lactate levels. Exercises such as the squat **(Figure 5-3)** will increase your strength as well as your anaerobic capacity. You will also continue to maintain and strengthen your training base.

Phase 2 Training Guidelines

a. Your cardiovascular workouts will reach zone 3.
b. Your strength workouts will see a decrease in repetitions to 8–10 per set.
c. CPAT–specific training intervals will increase in length and intensity to 90 seconds. Total interval time at a high intensity will be 5 minutes. This will accustom you to higher blood lactate levels.
d. You will start training the transitions between events.
e. Evaluate your progress by performing a goal check. Reset any unmet goals or adjust your training schedule, if necessary.
f. The last week of this phase is performed at a lower intensity of effort. It allows for physical recovery and enhancement of the adaptation process.

Figure 5-3 Building strength in phase 2.

> **Performance Point**
> Do not expect your performance to improve with each training session. It is not unusual to have a "down" day following a super training session. Your progress over a series of weeks is what counts.

Phase 3—Strength, Power, and Endurance

Objectives:

- Overlearn the CPAT events and transitions.
- Increase strength and power while maintaining endurance.
- Evaluate your phase 2 goals.

This phase promotes maximal strength development by using basic multi-joint exercises. Longer rest periods are used between strength training sets. Your increasing muscular strength, improving coordination, and efficiency of movement will develop power (the ability to use strength rapidly). Increasing your power production is crucial in passing the CPAT.

Phase 3 Training Guidelines

- **a.** Your cardiovascular workouts will involve zones 2, 3, and 4.
- **b.** CPAT–specific intervals will increase in length and intensity to 120 seconds. Total interval time at a high intensity will be 15 minutes. This will allow you to overlearn the skills and transitions involved in each event. Perfecting mental focus is also necessary here.
- **c.** Strength training will include endurance sets and power sets of 8-6-4 repetitions. Allow 2–4 minutes rest between sets. The key is to strengthen muscles through slow, controlled weight training and then translate the increased strength into power.
- **d.** Rest is crucial during this phase. Intensity is increasing and you must increase your rest days. Similar to the preceding phases, the last week of this phase is performed at a lower intensity of effort to allow for physical and mental recovery.
- **e.** Evaluate your progress by performing a goal check **(Figure 5-4)**. Adjust or reset any unmet goals and modify your training schedule, if necessary.

Figure 5-4 Evaluating your goals will keep you on track.

Phase 4—Peak Power

Objectives:

- Translate strength into power.
- Overlearn the CPAT events and transitions.
- Train for longer durations at your lactate threshold.
- Evaluate your phase 3 goals.

This phase involves peaking for the actual test. Success in this phase results largely from power production (the ability to use strength rapidly). You will focus on increasing your ability to generate power through interval training and lactate threshold training. This will give you the ability to train faster and longer, thus maintaining power through all eight events.

The strength gains you have made during the last two phases will enhance your CPAT–specific skill training. Increasing and maintaining muscular strength is a priority because as strength decreases so does performance.

During the final week of this phase, intensity will taper off with the last 2 days preceding the test consisting of only active rest (light walking, stretching, etc.).

Phase 4 Training Guidelines

- **a.** Cardiovascular workouts will consist of high-intensity interval training. Your heart rate will reach zones 4 and 5. Low-intensity cardio is included to maintain your training base and to prevent overtraining.
- **b.** CPAT–specific intervals will increase to 3 minutes in length and total interval workout time equaling about 30 minutes. This will allow you to overlearn the events and will accustom you to higher blood lactate levels.
- **c.** Strength workouts for the first 2 weeks in this phase will consist of two power sets of 6 repetitions per set. During the last 2 weeks of the phase, repetitions will increase to two sets of 10–12 reps. Primary exercises should be performed quickly and explosively two times per week. These are executed at a

high rate of speed to enhance and maintain power development.

d. Your transition training between events should be perfected by this phase. The movements should be part of your subconscious, allowing you to mentally focus on the next event.

e. Rest is crucial during this phase. Intensity is increasing and you must increase your rest days. The last week of this phase consists of lower intensity work and mental preparation. You should feel your body is getting stronger. By test day, you are rested, strong, and focused.

f. Evaluate your progress by performing a goal check. Adjust or reset any unmet goals. Adjust your training schedule, if necessary.

Performance Point

Working extra hard in the final days before the test does nothing but drain your energy. You want to be rested and mentally ready. Rest completely, except for light walking and stretching the last 2 days before the test. Rest your mind also. Use relaxation techniques and take your mind off the test.

PLANNING GUIDELINES FOR THE CPAT

If you wish to design your own program use the following guidelines:

1. Go through the biometric tests. This will provide you with baseline measures and allow you to set your goals for the entire program.
2. Divide the total time by 4. This will give you the number of weeks in each phase.
3. Plan the first phase (base training). Your starting intensities will be determined by the results from your biometric tests.
4. Plan the remaining phases using the principles described in the preceding section. As you progress through the phases, the intensity of training should increase by 10% to 20%.
5. Develop the last phase (peak power) by working backward from the date of the CPAT test. The 2 days preceding the test should be complete rest and recovery days. The last week should consist of low-intensity exercises. In the early part of the phase, your training intensity should be at the peak power level.
6. Evaluate the plan often. If you are meeting your goals and objectives, continue on to the next phase. If your objectives are not achieved, set new goals and redesign the next phase.
7. If you miss a week of training because of illness or injury, begin where you left off. Do not skip ahead. If necessary, adjust the time spent in the remaining phases to meet the CPAT test deadline.
8. Monitor your training progress and if overtraining becomes evident, modify the program. In the overall scheme, rest and recovery days are just as important as workout days.
9. If you only have a few weeks to train for the CPAT, concentrate on quality training in two phases: base training and peak power.

Performance Point

How much time can you realistically spend on training? Most people tend to overestimate the amount of time they can commit and this leads to frustration. Most people have many commitments (job, significant other, family, etc.). Training is less stressful when you find ways to include other people in your training process. For example, a significant other could record times and give encouragement during the training sessions.

CPAT TEST DAY

You have spent months preparing mentally and physically for the CPAT. What you do in the hours before, during, and after the event can also affect your experience.

Before the test:

- Make sure you are well rested. If traveling is necessary for the test, try to arrive a few days early to rest and acclimatize. Drink plenty of water as travel tends to dehydrate the system. Get a good night's sleep for several nights preceding the test. Chances are that anxiety and anticipation will lead to poor night's sleep the night before.
- Stay with your familiar nutrition program—do not introduce new foods. You do not want a case of gastric problems on test day.

- Scope out the test grounds ahead of time. Set yourself mentally for the arrangement of events. Make adjustments for your focus, if necessary.

Test day:

- Time the pre-test meal according to the way you ate meals during training. Stay hydrated. A common mistake is to skip the pre-test meal and drink little to avoid frequent bathroom trips.
- Stay focused. Enjoy the day and let it unfold one step at a time. Maintain concentration—don't get caught up in the crowd and what other people are doing.

After the test:

- Change your clothes, stretch, or get a massage. Relax and enjoy the moment.
- Rehydrate and refuel.
- Don't be alarmed if you experience "post-test depression." Many people liken the feeling to the depression experienced by new mothers.
- Relax and enjoy activities with friends.

INJURIES

Whether a nagging ache or a more serious problem, injuries do not necessarily mean an end to preparing for the CPAT. Most athletes experience some type of injury during their training schedules and training for the CPAT is no different. Proper diagnosis and treatment of injuries are critical to maintaining a training schedule.

Delayed-Onset Muscle Soreness

Muscle soreness that occurs 24 to 48 hours after an activity is called **delayed-onset muscle soreness (DOMS)**. This usually occurs when starting a new activity and is a sign that the body is adapting to a new challenge. The soreness should subside within a few days. The best strategy is to perform active rest (a light activity with the body parts involved), take a nonsteroidal anti-inflammatory drug (NSAID), and stretch gently after exercise. The CPAT periodization model has built in rest periods to allow for recovery from DOMS.

Acute Injuries

Injuries involving pain lasting longer than a week (e.g., redness, shooting pain, swelling, or a clicking sound in the joints) may be an indication of a severe injury that requires a doctor's attention. Acute injuries can be the product of blunt trauma or overstretching a muscle, joint, or tendon. Overstretching can cause a tear, sprain, or strain, and if left untreated may turn into a chronic condition.

Chronic Injuries

Chronic injuries are often the result of overtraining. Repeated bouts of high stress, sudden increases in activity, running on new or uneven surfaces, and inadequate shoe support can cause injuries such as shin splints, plantar faciitis, and Achilles tendonitis.

Treating Injuries

The traditional treatment for these injuries entails using **RICE**—an acronym for rest, ice, compression, and elevation **(Figure 5-5)**. As we learn more about the body's own healing mechanism, RICE may not be the best treatment. The latest research indicates that immobilizing an injury (unless it is a fracture, torn muscle, or your doctor advises you to) restricts blood flow and decreases the temperature of the tissues which needs to be at 100.7°F for the cells to do their job.

Treating injuries by using the **MICE** technique moves the affected area through its functional range of motion. This increases blood flow to the area, maintaining a healing temperature for the tissues. Guidelines for using MICE are as follows:

Movement—Slowly move the affected area through its range of motion. Use gentle stretching.
Ice—Put ice on the injury for 20 minutes, 3 to 4 times per day. Place a towel between the skin and the ice so you don't get frostbite. A bag of frozen corn or peas makes a great reusable ice bag.
Compression—Wrap a towel or Ace-type bandage around the area. It should be tight enough to support the joint but not too tight. Start wrapping away from the heart and apply the bandage upward. Be cautious about sleeping while wearing something that can compromise circulation.
Elevation—Raise the injured body part so it is slightly higher than your heart.

Performance Point

Injuries rarely "just happen." Listen to your body. Pay attention to warning signs: pain, soreness, aching, and fatigue. Never brush off suspicious symptoms. Instead, make immediate adjustments to your training or seek a professional opinion.

Figure 5-5 Using RICE to treat injuries.

OVERTRAINING

If training for the CPAT is leaving you more exhausted than energized, overtraining is a possibility. **Overtraining** is decreased work capacity resulting from too little rest or too much training. Excessive training can put your health at risk, as you will break down more than you build up.

Symptoms of overtraining include:

- Decreased performance
- Agitation, moodiness, irritability or lack of concentration
- Excessive fatigue
- Increased perceived effort during workouts
- Chronic or nagging muscle aches or joint pain
- Frequent illnesses and upper respiratory infections
- Insomnia or restless sleep
- Loss of appetite and weight loss
- Elevated resting heart rate
- Menstrual cycle disturbances in women
- Muscle or joint pain that lingers longer than 48 hours

Guidelines for Overtraining

1. Make sure you are getting the nutrients to support your training. Eat within 15 to 30 minutes after each training session. Adequate amounts of complex carbohydrates are essential for the increased demands placed on your system.
2. Take at least 1 rest day per week and additional days if needed. Resume the program where you left off. Do not skip ahead a week to make up for lost time. If you miss 3 to 4 weeks, start back at the base training phase. You will need to recondition your joints and the affected muscle area.
3. Check your pulse for 60 seconds before getting out of bed. If it is 5 beats higher than normal, you are due for a rest day.
4. Increase your sleep time. This is when recuperation and growth take place.
5. Post-workout ice baths and sports massages improve circulation and flush out waste products, reducing inflammation and soreness.

Case Studies: Using Goals to Achieve Success

Candidates Angie and Bill developed the following training goals after taking a trial run on the CPAT course.

Case 1

Angie set two specific goals:

Goal 1—Decrease her time from 16:11 to 9:45.

Objectives:

a. Break down her time at each event and between events.
b. Analyze her technique at each event.
c. Develop a mental focus for each event.

Goal 2—Learn the CPAT events and develop a strategy for each.

Objectives:

a. View a tape of a successful candidate.
b. Interview three successful candidates (one woman).
c. Enroll in a CPAT preparation course at a community college.
d. Practice CPAT specific drill twice a week.

Angie enrolled in a CPAT preparation course at the local community college. The 12-week program consisted of a 12-week periodized plan with the final day being the actual CPAT test. The course included CPAT specific drills, weight training, and cardiovascular and core body exercises. Classroom work involved goal setting, strategies for each of the eight events, and the development of mental focusing techniques. At the 8-week point, Angie took a second CPAT trial run. Her time dropped to 12:30, an improvement of 3:41. She analyzed her time at each station and transition time between events. She also received feedback from fellow students and staff. From this, she set new goals and developed several new focus points for the events. One of these involved her weakest event, the Rescue (dummy drag). A staff member observed her rounding her back about halfway through the event, which resulted in her stopping two times before crossing the finish line. She developed the mantra "strong legs, strong back," and visualized herself completing the event without stopping. The staff member also noticed that she was using her arms to swing the hammer in the Forcible Entry event. Angie visualized power from her lower body being transferred into the hammer swing.

The last 4 weeks of the course focused on generating power in the CPAT drills and weight training exercises. Angie felt herself getting stronger. She now could drag the dummy 90 ft without stopping and complete the CPAT Forcible Entry drill with increased power and perfect technique. She fine-tuned her focus points for each of the events and visualized herself completing each event. During the final week she took 2 days off preceding the test, relaxed with her family, and took her mind off the test.

When the CPAT test day arrived, she was rested, mentally confident, and focused. Her result is listed below.

Angie passed the test with 38 seconds to spare. She celebrated with her family that evening.

Event No.	Time (seconds)	Transition time (seconds)
1. Step Mill	180	24
2. Hose Drag	30	22
3. Equipment Carry	38	18
4. Ladder Raise/Extn	20	19
5. Forcible Entry	31	20
6. Search	46	21
7. Rescue	41	22
8. Ceiling Breach/Pull	50	—
Total (seconds)	436	146
Total time (minutes)	9:42 (pass)	

Case Studies: Using Goals to Achieve Success—Cont'd

Case 2

Bill set two specific goals:

Goal 1—Learn the CPAT events and develop a mental focus for each event.

Objectives:

a. View a video tape of the CPAT.
b. Develop a CPAT circuit and practice CPAT specific drills twice a week.

Goal 2—Develop a training plan which will allow him to "peak" for the CPAT.

Objectives:

a. Have a nationally certified personal trainer develop a periodized training plan for him.
b. Train with another candidate.

Bill continued his weight training program but modified it according to the personal trainer's recommendation. The new program allowed Bill to recuperate after intense training sessions. The trainer also substituted treadmill walking on an incline instead of stair running. Bill noticed his back and knees stopped aching and his energy level increased.

Bill's wife acted as timer when he and his training partner performed their CPAT circuits. Bill visualized himself on each station of the real CPAT, performing with flawless technique and perfect execution.

Flexibility training was an important part of his new program. While performing the stretches, Bill and his partner reviewed the workout just completed. They provided each other feedback about their performance. Bill felt confident when test day arrived. He now had a plan and focus points for each of the events. His result is listed below.

Bill passed the test with 1:15 seconds to spare. His wife treated him to dinner that evening.

Event No.	Time (seconds)	Transition time (seconds)
1. Step Mill	180	22
2. Hose Drag	28	22
3. Equipment Carry	36	18
4. Ladder Raise/Extn	20	19
5. Forcible Entry	15	17
6. Search	45	21
7. Rescue	32	22
8. Ceiling Breach/Pull	48	—
Total (seconds)	404	141
Total time (minutes)	9:05 (pass)	

CHAPTER SUMMARY

Having a carefully designed training program is crucial to succeeding in the CPAT.

Periodization is a training program that seeks to "peak" an athletes performance with a competitive event. This is accomplished by changing training intensity and strategically placing maintenance and recovery phases to enhance performance.

Periodized training programs are based on the general adaptation syndrome which states that properly timed stress, or eustress, will produce a supercompensation effect and thus improve performance. Periodized programs have shown to reduce injuries, prevent overtraining, and produce better results than nonperiodized ones.

The CPAT periodization model consists of four phases of increasing intensity.

Phase 1 is base training which builds a foundation for the higher intensity training found in the following phases. The CPAT events are learned and fitness testing is done to determine starting points for all exercises.

Phase 2 focuses on developing strength and endurance by increasing training intensity. Candidates will also start training the transitions between events and start to develop mental focusing techniques.

Phase 3 integrates strength and endurance to produce power (the ability to apply strength rapidly). This enables the CPAT candidate to maintain performance for the duration of the test.

Phase 4 focuses on developing peak power for the actual CPAT test. CPAT events and transitions are overlearned to ensure that performance is efficient and automatic.

Goals are evaluated and reset if necessary at the end of each phase.

Being well rested, staying with a familiar nutrition plan, and becoming familiar with the test layout can increase chances for success.

Obstacles such as muscle soreness, injuries, and overtraining sometimes do appear but they do not necessarily mean an end to training if handled correctly.

Muscle soreness that usually subsides within 2 days after undertaking a new activity is called delayed onset muscle soreness. It is not a serious condition and is best treated by low-intensity activity of the affected areas. Injuries requiring a doctor's attention include acute injuries which can be the result of overstretching or blunt trauma, and chronic injuries which usually result from overtraining. The recommended method of treating injuries is called MICE (gentle movement, ice, compression, and elevation), which enhances the body's natural healing mechanisms.

Overtraining is best treated by increasing rest days and ensuring nutrition is adequate to handle the increased demands of training. Monitoring of the training program will reduce the likelihood of overtraining and injuries.

CHECK YOUR LEARNING

1. The concept of supercompensation refers to
- **a.** Working hard right up to test day.
- **b.** Adapting to training demands.
- **c.** Having someone train with you.
- **d.** Increasing your energy by "carb loading."

2. Benefits of periodization include
- **a.** Enabling you to peak for the CPAT.
- **b.** Planning out training.
- **c.** Avoiding injuries.
- **d.** All of the above.

3. Phase 1 training involves
- **a.** High-intensity training.
- **b.** Developing power.
- **c.** Perfecting skills.
- **d.** None of the above.

4. All of these are components of the general adaptation syndrome *except*
- **a.** Eustress.
- **b.** Maintenance.
- **c.** Supercompensation.
- **d.** Distress.

5. A good treatment for overtraining is to
- **a.** Increase your protein intake and decrease your carbohydrate intake.
- **b.** Take hot showers.
- **c.** Get post-workout massages.
- **d.** Decrease your rest days.

6. Phase 4 training involves
- **a.** Developing peak power.
- **b.** Perfecting skills.
- **c.** Using active rest.
- **d.** All of the above.

7. DOMS is caused by
- **a.** Overstretching a muscle.
- **b.** Warming up too long.
- **c.** Keeping training intensity constant.
- **d.** Starting new training methods.

8. Phase 2 training involves
- **a.** Limited rest time.
- **b.** Focusing on strength over endurance.
- **c.** Developing mental focus.
- **d.** Decreasing carbohydrate intake.

9. The concept of power refers to
- **a.** The ability to lift a great amount of weight.
- **b.** The ability to lift a heavy weight several times.
- **c.** The ability to train for extended times.
- **d.** The ability to use strength rapidly.

10. If your preparation time is limited, which two training phases should you focus on?
- **a.** Phases 1 and 3.
- **b.** Phases 1 and 4.
- **c.** Phases 2 and 4.
- **d.** Phases 2 and 3.

References

1. V.M. Zatsiorsky, *Science and Practice of Strength Training* (Champaign, IL: Human Kinetics, 1995).
2. T.O. Bompa, *Periodization: Theory and Methodology of Training* (Champaign, IL: Human Kinetics, 1999).
3. A. Lydiard, G. Gilmour, *Running the Lydiard Way* (Mountain View, CA: World Publications, 1978). Cited in T. D. Noakes, *Love of Running* (Champaign, IL: Leisure Press, 1991, pp. 155, 157, 209).
4. D.S. Willoughby, "The Effects of Mesocycle-length Weight Training Programs Involving Periodization and Partially Equated Volumes on Upper and Lower Body Strength," *Journal of Strength and Conditioning Research*, vol. 7, pp. 2–8, 1993.

WEIGHT TRAINING EXERCISES

CHEST EXERCISES

A Dumbbell press

B Dumbbell press

A Barbell press

B Barbell press

CHEST EXERCISES

A Incline dumbbell press

B Incline dumbbell press

A Incline barbell press

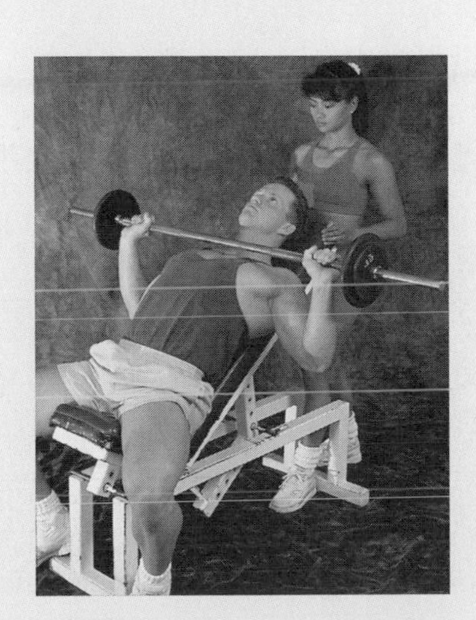

B Incline barbell press

BACK EXERCISES

A Barbell bent over row

B Barbell bent over row

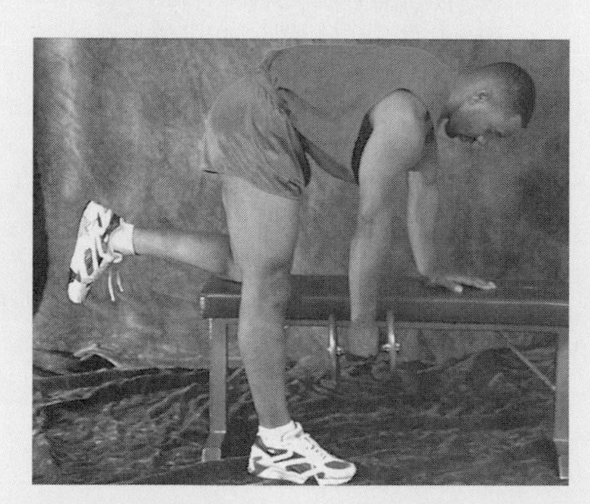

A One-dumbbell bent over row

B One-dumbbell bent over row

BACK EXERCISES

A Reverse grip chin-up

B Reverse grip chin-up

A Forward grip pull-up

B Forward grip pull-up

BACK AND SHOULDER EXERCISES

A Lat pull downs

B Lat pull downs

A Dumbbell overhead press

B Dumbbell overhead press

SHOULDER AND BICEP EXERCISES

A Barbell overhead press

B Barbell overhead press

A Seated dumbbell curl

B Seated dumbbell curl

BICEP AND FOREARM EXERCISES

A Barbell curl

B Barbell curl

A Reverse curl

B Reverse curl

BICEP AND TRICEP EXERCISES

A Incline dumbbell curl

B Incline dumbbell curl

A Tricep pushdowns

B Tricep pushdowns

TRICEP EXERCISES

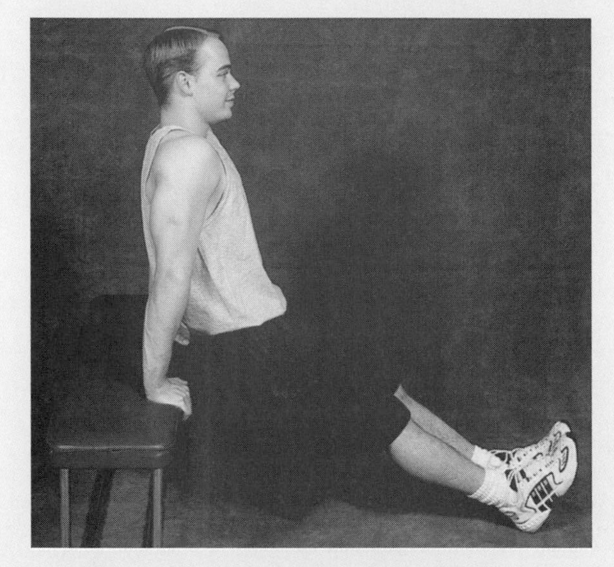

A Bench dips

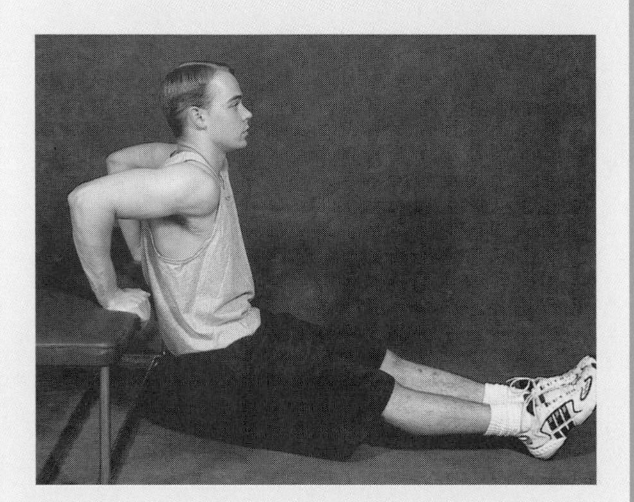

B Bench dips

A Close grip bench press

B Close grip bench press

LEG AND BACK EXERCISES

A Barbell squat

B Barbell squat

A Deadlift

B Deadlift

LEG EXERCISES

A Barbell squat

B Barbell squat

A Lunge

B Lunge

LEG EXERCISES

A Step up with dumbbells

B Step up with dumbbells

A Leg press

B Leg press

LEG EXERCISES

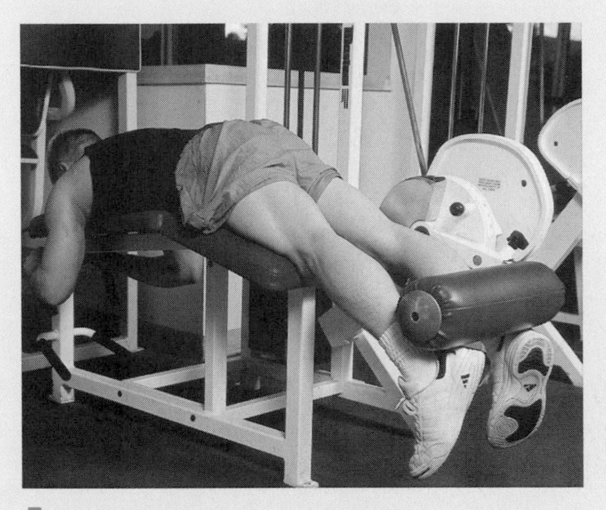

A Leg curl

B Leg curl

A Dumbbell calf raise

B Dumbbell calf raise

ABDOMINAL EXERCISES

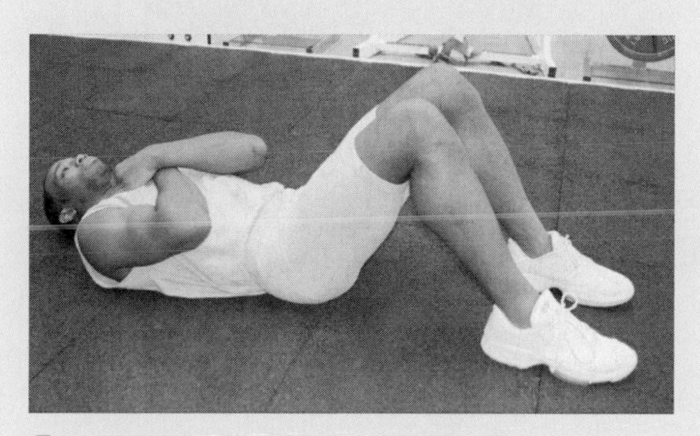

A Crunches

B Crunches

A Reverse crunches

B Reverse crunches

ABDOMINAL EXERCISES

A Hanging knee raises

B Hanging knee raises

TRAINING SCHEDULES

Listed on the following pages are 16, 12, 8, and 4 week training plans. Choose the plan which best fits into your time schedule.

16-WEEK SEMESTER PLAN

Phase 1—Base Training

WEEK 1	MONDAY	TUESDAY	THURSDAY	FRIDAY	SATURDAY
Classroom	Introduction to the CPAT The four success principles		Goal Setting		
Workout	Module 1—Biometric testing	Weight training (1A) Cardio training (1A) Flexibility	Functional skills (1A) CPAT specific drill demo Flexibility	Weight training (1B) Cardio training (1B) Flexibility	CPAT trial course

WEEK 2	MONDAY	TUESDAY	THURSDAY	FRIDAY	SATURDAY
Classroom	Module 2—Nutritional Concepts Module 3—Fitness Training Principles		Module 4—Mental Training		
Workout	Functional skills (1A) CPAT drills (1A) Flexibility	Weight training (1A) Cardio training (1A) Flexibility	Functional skills (1A) CPAT drills (1B) Flexibility	Weight training (1B) Flexibility	Cardio training (1B) Flexibility

WEEK 3	MONDAY	TUESDAY	THURSDAY	FRIDAY	SATURDAY
Classroom	Module 5—Stair Climb Module 6—Hose Drag		Module 7—Equipment Carry Module 8—Ladder Raise & Extension		
Workout	Functional skills (1A) CPAT drills (1A) Flexibility	Weight training (1A) Cardio training (1A) Flexibility	Functional skills (1A) CPAT drills (1B) Flexibility	Weight training (1B) Flexibility	Cardio training (1B) Flexibility

WEEK 4	MONDAY	TUESDAY	THURSDAY	FRIDAY	SATURDAY
Classroom	Module 9—Forcible Entry Module 10—Search		Module 11—Rescue Module 12—Ceiling Breach & Pull		
Workout	Functional skills (1A) CPAT drills (1A) Flexibility	Weight training (1A) Cardio training (1A) Flexibility	Functional skills (1A) CPAT drills (1B) Flexibility	Weight training (1B) Flexibility	Cardio training (1B) Flexibility

16-WEEK SEMESTER PLAN

Phase 2—Strength and Endurance

WEEK 5	MONDAY	TUESDAY	THURSDAY	FRIDAY	SATURDAY
Classroom					
Workout	Functional skills (2B) CPAT drills (2A) Flexibility	Weight training (2A) Cardio training (2A) Flexibility	Functional skills (2B) CPAT drills (2B) Flexibility	Weight training (2B) Flexibility	Cardio training (2B) Flexibility

WEEK 6	MONDAY	TUESDAY	THURSDAY	FRIDAY	SATURDAY
Classroom	Goal check				
Workout	Functional skills (2B) CPAT drills (1A) Flexibility	Weight training (2A) Cardio training (2A) Flexibility	Functional skills (2A) CPAT drills (2B) Flexibility	Weight training (2B) Flexibility	Cardio training (2B) Flexibility

WEEK 7	MONDAY	TUESDAY	THURSDAY	FRIDAY	SATURDAY
Classroom					
Workout	Functional skills (2B) CPAT drills (2A) Flexibility	Weight training (2A) Cardio training (2A) Flexibility	Functional skills (2A) CPAT drills (2B) Flexibility	Weight training (2B) Flexibility	Cardio training (2B) Flexibility

WEEK 8	MONDAY	TUESDAY	THURSDAY	FRIDAY	SATURDAY
Classroom					
Workout	2nd Biometric testing	Weight training (2A) Cardio training (2A) Flexibility	Functional skills (2A) CPAT drills (2B) Flexibility	Weight training (2B) Flexibility	Cardio training (2B) Flexibility

16-WEEK SEMESTER PLAN

Phase 3—Strength, Power, and Endurance

WEEK 9	MONDAY	TUESDAY	THURSDAY	FRIDAY	SATURDAY
Classroom	Goal check				
Workout	Functional skills (3B) CPAT drills (3A) Flexibility	Weight training (3A) Cardio training (3A) Flexibility	Functional skills (3A) CPAT drills (3B) Flexibility	Weight training (3B) Flexibility	Cardio training (3B) Flexibility

WEEK 10	MONDAY	TUESDAY	THURSDAY	FRIDAY	SATURDAY
Classroom					
Workout	Functional skills (3B) CPAT drills (3A) Flexibility	Weight training (3A) Cardio training (3A) Flexibility	Functional skills (3A) CPAT drills (3B) Flexibility	Weight training (3B) Flexibility	Cardio training (3B) Flexibility

WEEK 11	MONDAY	TUESDAY	THURSDAY	FRIDAY	SATURDAY
Classroom					
Workout	Functional skills (3B) CPAT drills (3A) Flexibility	Weight training (3A) Cardio training (3A) Flexibility	Functional skills (3A) CPAT drills (3B) Flexibility	Weight training (3B) Flexibility	Cardio training (3B) Flexibility

WEEK 12	MONDAY	TUESDAY	THURSDAY	FRIDAY	SATURDAY
Classroom					
Workout	3rd Biometric testing	Weight training (3A) Cardio training (3A) Flexibility	Functional skills (3A) CPAT drills (3B) Flexibility	Weight training (3B) Flexibility	Cardio training (3B) Flexibility

16-WEEK SEMESTER PLAN

Phase 4—Peak Power

WEEK 13	MONDAY	TUESDAY	THURSDAY	FRIDAY	SATURDAY
Classroom	Goal check				
Workout	Functional skills (4B) CPAT drills (4A) Flexibility	Weight training (4A) Cardio training (4A) Flexibility	Functional skills (4A) CPAT drills (4B) Flexibility	Weight training (4B) Flexibility	Cardio training (4B) Flexibility

WEEK 14	MONDAY	TUESDAY	THURSDAY	FRIDAY	SATURDAY
Classroom					
Workout	Functional skills (4B) CPAT drills (4A) Flexibility	Weight training (4A) Cardio training (4A) Flexibility	Functional skills (4A) CPAT drills (4B) Flexibility	Weight training (4B) Flexibility	Cardio training (4B) Flexibility

WEEK 15	MONDAY	TUESDAY	THURSDAY	FRIDAY	SATURDAY
Classroom					
Workout	Functional skills (4B) CPAT drills (4A) Flexibility	Weight training (4A) Cardio training (4A) Flexibility	Functional skills (4A) CPAT drills (4B) Flexibility	Weight training (4B) Flexibility	Cardio training (4B) Flexibility

WEEK 16	MONDAY	TUESDAY	THURSDAY	FRIDAY	SATURDAY
Classroom	Goal check	Mental rehearsal of CPAT course			
Workout	Functional skills (4A) CPAT drills (4A) Flexibility	Weight training (4B) Cardio training (4B) Flexibility	Light walking Stretching	Light walking Stretching	CPAT test day

12-WEEK PLAN

Phase 1—Base Training

WEEK 1	MONDAY	TUESDAY	THURSDAY	FRIDAY	SATURDAY
Classroom	Introduction to the CPAT	Goal setting	CPAT trial course	The Four Success Principles	
Workout	Module 1—Biometric Testing	Weight training (1A) Cardio training (1A) Flexibility	Functional skills (1A) CPAT specific drill demo Flexibility	Weight training (1B) Cardio training (1B) Flexibility	

WEEK 2	MONDAY	TUESDAY	THURSDAY	FRIDAY	SATURDAY
Classroom	Module 2—Nutritional Concepts		Module 3—Fitness Training Principles		
Workout	Functional skills (1B) CPAT drills (1A) Flexibility	Weight training (1A) Cardio training (1A) Flexibility	Functional skills (1A) CPAT drills (1B) Flexibility	Weight training (1B) Flexibility	Cardio training (1B) Flexibility

WEEK 3	MONDAY	TUESDAY	THURSDAY	FRIDAY	SATURDAY
Classroom	Module 4—Mental Training		Module 5—Stair Climb Module 6—Hose drag		
Workout	Functional skills (1B) CPAT drills (1A) Flexibility	Weight training (1A) Cardio training (1A) Flexibility	Functional skills (1A) CPAT drills (1B) Flexibility	Weight training (1B) Flexibility	Cardio training (1B) Flexibility

WEEK 4	MONDAY	TUESDAY	THURSDAY	FRIDAY	SATURDAY
Classroom	Module 7—Equipment Carry Module 8—Ladder Raise & Extension		Module 9—Forcible Entry Module 10—Search		
Workout	Functional skills (1B) CPAT drills (1A) Flexibility	Weight training (1A) Cardio training (1A) Flexibility	Functional skills (1A) CPAT drills (1B) Flexibility	Weight training (1B) Flexibility	Cardio training (1B) Flexibility

12-WEEK PLAN

Phase 2—Strength and Endurance

WEEK 5	MONDAY	TUESDAY	THURSDAY	FRIDAY	SATURDAY
Classroom	Module 11—Rescue Module 12—Ceiling Breach & Pull				
Workout	Functional skills (2B) CPAT drills (2A) Flexibility	Weight training (2A) Cardio training (2A) Flexibility	Functional skills (2A) CPAT drills (2B) Flexibility	Weight training (2B) Flexibility	Cardio training (2B) Flexibility

WEEK 6	MONDAY	TUESDAY	THURSDAY	FRIDAY	SATURDAY
Classroom					
Workout	Functional Skills (2B) CPAT drills (2A) Flexibility	Weight training (2A) Cardio training (2A) Flexibility	Functional skills (2A) CPAT drills (2B) Flexibility	Weight training (2B) Flexibility	Cardio training (2B) Flexibility

WEEK 7	MONDAY	TUESDAY	THURSDAY	FRIDAY	SATURDAY
Classroom					
Workout	Functional skills (3B) CPAT drills (3A) Flexibility	Weight training (3A) Cardio training (3A) Flexibility	Functional skills (3A) CPAT drills (3B) Flexibility	Weight training (3B) Flexibility	Cardio training (3B) Flexibility

WEEK 8	MONDAY	TUESDAY	THURSDAY	FRIDAY	SATURDAY
Classroom	Goal check				
Workout	2nd Biometric testing	Weight training (3A) Cardio training (3A) Flexibility	Functional skills (3A) CPAT drills (3B) Flexibility	Weight training (3B) Flexibility	Cardio training (3B) Flexibility

12-WEEK PLAN

Phase 3—Peak Power

WEEK 9	MONDAY	TUESDAY	THURSDAY	FRIDAY	SATURDAY
Classroom					
Workout	Functional skills (4B) CPAT drills (4A) Flexibility	Weight training (4A) Cardio training (4A) Flexibility	Functional skills (4A) CPAT drills (4B) Flexibility	Weight training (4B) Flexibility	Cardio training (4B) Flexibility

WEEK 10	MONDAY	TUESDAY	THURSDAY	FRIDAY	SATURDAY
Classroom					
Workout	Functional skills (4B) CPAT drills (4A) Flexibility	Weight training (4A) Cardio training (4A) Flexibility	Functional skills (4A) CPAT drills (4B) Flexibility	Weight training (4B) Flexibility	Cardio training (4B) Flexibility

WEEK 11	MONDAY	TUESDAY	THURSDAY	FRIDAY	SATURDAY
Classroom	3rd Biometric testing	Goal check			
Workout	Functional skills (4B) CPAT drills (4A) Flexibility	Weight training (4A) Cardio training (4A) Flexibility	Functional skills (4A) CPAT drills (4B) Flexibility	Weight training (4B) Flexibility	Cardio training (4B) Flexibility

WEEK 12	MONDAY	TUESDAY	THURSDAY	FRIDAY	SATURDAY
Classroom	Mental rehearsal of CPAT course				
Workout	Functional skills (4B) CPAT drills (4A) Flexibility	Weight training (4B) Cardio training (4B) Flexibility	Light walking Stretching	Light walking Stretching	CPAT test day

8-WEEK PLAN

Phase 1—Base Training and Endurance

WEEK 1	MONDAY	TUESDAY	THURSDAY	FRIDAY	SATURDAY
Study	Introduction to the CPAT The Four Success Priciples	Goal setting Chapter 3	Chapter 4 Modules 5–6	Chapter 4 Modules 7–8	CPAT trial course
Workout	1st Biometric testing	Weight training (1A) Cardio training (1A) Flexibility	Functional skills (1A) CPAT drills (1A) Flexibility	Weight training (1B) Cardio training (1B) Flexibility	

WEEK 2	MONDAY	TUESDAY	THURSDAY	FRIDAY	SATURDAY
Study	Chapter 4	Chapter 4	Modules 9–10	Modules 11–12	
Workout	Functional skills (1B) CPAT drills (1A) Flexibility	Weight training (1A) Cardio training (1A) Flexibility	Functional skills (1A) CPAT drills (1B) Flexibility	Weight training (1B) Flexibility	Cardio training (1B) Flexibility

WEEK 3	MONDAY	TUESDAY	THURSDAY	FRIDAY	SATURDAY
Study					
Workout	Functional skills (2B) CPAT drills (2A) Flexibility	Weight training (2A) Cardio training (2A) Flexibility	Functional skills (2A) CPAT drills (2A) Flexibility	Weight training (2B) Flexibility	Cardio training (2B) Flexibility

WEEK 4	MONDAY	TUESDAY	THURSDAY	FRIDAY	SATURDAY
Study					
Workout	Functional skills (2B) CPAT drills (2A) Flexibility	Weight training (2A) Cardio training (2A) Flexibility	Functional skills (2A) CPAT drills (2B) Flexibility	Weight training (2B) Flexibility	Cardio training (2B) Flexibility

8-WEEK PLAN

Phase 2—Strength and Peak Power

WEEK 5	MONDAY	TUESDAY	THURSDAY	FRIDAY	SATURDAY
Study	2nd Biometric testing Goal check				
Workout	Functional skills (3B) CPAT drills (3A) Flexibility	Weight training (3A) Cardio training (3A) Flexibility	Functional skills (3A) CPAT drills (3B) Flexibility	Weight training (3B) Flexibility	Cardio training (3B) Flexibility

WEEK 6	MONDAY	TUESDAY	THURSDAY	FRIDAY	SATURDAY
Study					
Workout	Functional skills (3B) CPAT drills (3A) Flexibility	Weight training (3A) Cardio training (3A) Flexibility	Functional skills (3A) CPAT drills (3B) Flexibility	Weight training (3B) Flexibility	Cardio training (3B) Flexibility

WEEK 7	MONDAY	TUESDAY	THURSDAY	FRIDAY	SATURDAY
Study					
Workout	Functional skills (4B) CPAT drills (4A) Flexibility	Weight training (4A) Cardio training (4A) Flexibility	Functional skills (4A) CPAT drills (4B) Flexibility	Weight training (4B) Flexibility	Cardio training (4B) Flexibility

WEEK 8	MONDAY	TUESDAY	THURSDAY	FRIDAY	SATURDAY
Study	Mental rehearsal of CPAT course				
Workout	Functional skills (4B) CPAT drills (4A) Flexibility	Weight training (4B) Cardio training (4B) Flexibility	Light walking Stretching	Light walking Stretching	CPAT test day

4-WEEK PLAN

Base Training and Peak Power

Note: students should read and be familiar with chapters 1–4 before implementing this schedule.

WEEK 1	MONDAY	TUESDAY	THURSDAY	FRIDAY	SATURDAY
Study	Biometric testing	Goal setting			
Workout	Functional skills (1B) CPAT drills (1A) Flexibility	Weight training (1A) Cardio training (1A) Flexibility	Functional skills (1A) CPAT drills (1B) Flexibility	Weight training (1B) Flexibility	Cardio training (1B) Flexibility

WEEK 2	MONDAY	TUESDAY	THURSDAY	FRIDAY	SATURDAY
Study					
Workout	Functional skills (2B) CPAT drills (2A) Flexibility	Weight training (1B) Cardio training (1B) Flexibility	Functional skills (2A) CPAT drills (2A) Flexibility	Weight training (2A) Flexibility	Cardio training (2A) Flexibility

WEEK 3	MONDAY	TUESDAY	THURSDAY	FRIDAY	SATURDAY
Study	Goal check				
Workout	Functional skills (3A) CPAT drills (3A) Flexibility	Weight training (3A) Cardio training (3A) Flexibility	Functional skills (4A) CPAT drills (4B) Flexibility	Weight training (4A) Flexibility	Cardio training (4A) Flexibility

WEEK 4	MONDAY	TUESDAY	THURSDAY	FRIDAY	SATURDAY
Study	Mental rehearsal of CPAT course				
Workout	Functional skills (4B) CPAT drills (4A) Flexibility	Weight training (4B) Cardio training (4B) Flexibility	Light walking, Stretching	Light walking, Stretching	CPAT test day

NUTRITION TRACKING JOURNALS

Make copies of the following charts to track your nutrition plan.

WEEKLY NUTRITION TRACKING JOURNAL

TRAINING PHASE:

Week: **Dates:**

	Day 1	Day 2	Day 3	Day 4	Day 5	Day 6	Day 7	Total calories per food group	Weekly % of total calories
Carbohydrates									
Complex									
Simple									
Protein									
Fats									
Good Fats									
Bad Fats									
Total Daily Calories									
Daily:									
% Carbs									
% Protein									
% Fats									

DAILY NUTRITION TRACKING JOURNAL

Goals: % total calories
Carbohydrate:
Protein:
Fats:

Converting grams to calories:
Caloric multiplier:
1 gram of protein = 4 calories
1 gram of fat = 9 calories
1 gram of protein = 4 calories

	Meal 1	Meal 2	Meal 3	Meal 4	Meal 5	Meal 6	Total calories per food group	Daily % of total calories
Carbohydrates								
Complex								
Simple								
Protein								
Fats								
Good Fats								
Bad Fats								
Total Daily Calories								

For each meal, determine the amount of calories from carbohydrates, proteins and fats. Then determine your daily intake for each of these categories and transfer the amounts to the weekly tracking journal.

EXAMPLE OF THE COURSE LAYOUT

Each testing center lays out its course differently. As long as the guidelines are followed, as stated in the *CPAT Manual,* the course can be laid out in any manner that will fit the available space. Every course will have the events in the same order and there will be 85-ft (25.91 m) distance between each event. The following is an example of a possible course layout.

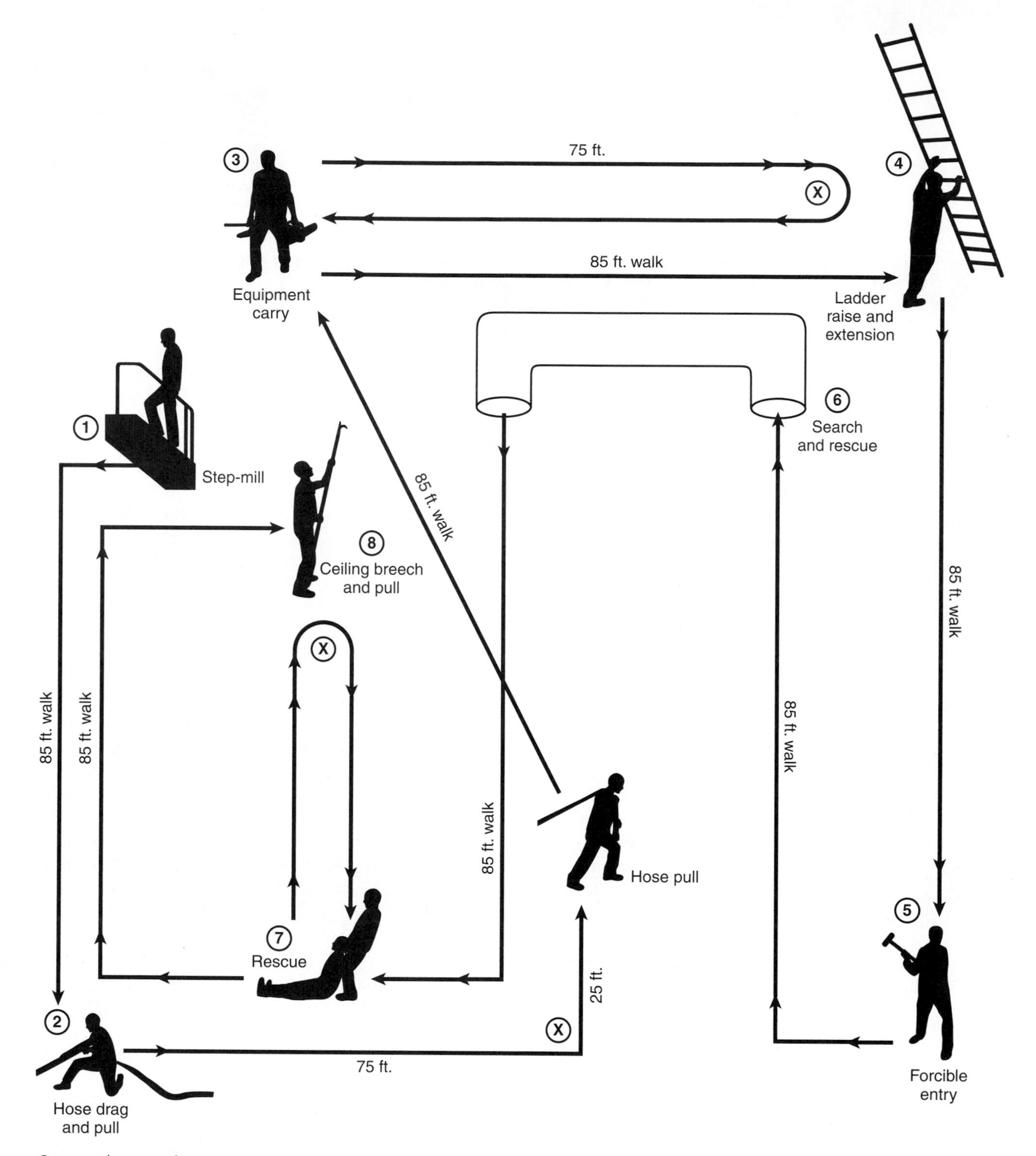

Course layout diagram.

GLOSSARY

Active practice Active practice provides as much experience as possible with the actual or simulated CPAT course.

Adopt To take and use as one's own.

Aerobic system The aerobic system uses oxygen primarily for low-intensity training and physical activities lasting longer than 3 minutes. The aerobic system builds the foundation for the high-intensity anaerobic system.

Anaerobic system The primary energy source for the anaerobic system is glucose which is developed from carbohydrates. This system produces a high amount of energy and causes the accumulation of waste products in the muscles and blood, such as lactic acid.

Apparatus The fire apparatus or fire truck.

Balance Stability or physical equilibrium.

Body composition Percentage of lean and fat tissue in the body.

BOSU balance trainer A dome-like platform with one flat side and one rounded soft side.

Candidate Physical Ability Test (CPAT) A physical test created to evaluate a new candidate's readiness to meet the high physical demands of a firefighting career.

Carbohydrates (complex) Complex carbohydrates are nutrients that are digested slowly and that provide energy for longer periods. Among all the nutrients, they are the most powerful in affecting energy levels.

Carbohydrates (simple) Simple carbohydrates are easily digested nutrients that quickly restore glycogen stores in the muscles. Simple carbohydrates such as energy drinks and gels can supply a quick source of energy.

CPAT Manual A manual that outlines the procedures to follow when administering the CPAT.

CPAT Preparation Guide Training Guide designed to ensure that firefighter candidates have the physical ability to perform the critical tasks of a firefighter effectively and safely. The CPAT is a pass/fail test that consists of eight (8) separate events.

Centering A mental training skill that involves breathing and focusing to position yourself for optimal performance.

Critical incident stress debriefing programs Informal counseling services to assist firefighters after traumatic incidents.

Delayed-onset muscle soreness (DOMS) Delayed-onset muscle soreness is muscle soreness that occurs 24 to 48 hours after an activity.

Distress A negative adaptation to stress, resulting when there is too great a stimulus and/or too little regeneration.

Dual-energy x-ray scan (DEXA) A way to measure body composition that uses a whole body scanner with low dose x-rays to read bone and tissue mass. DEXA analysis shows where fat is distributed throughout the body.

Employee assistance programs Formal counseling programs that cover a range of issues that may be of concern to employees.

Eustress A positive adaptation to stress, resulting from correctly timed alternation between stress and regeneration.

Fats Fats carry and store fat-soluble vitamins, construct cell membranes, and play a role in the production of testosterone and estrogen. The two types of naturally occurring fats are: saturated fats and unsaturated fats.

Flexibility Flexibility is defined as the capacity to move freely in every intended direction.

Fly section The fly section is the smaller, movable section of an extension ladder.

Focusing A mental training skill that involves excluding all irrelevant thoughts and emotions.

Functional skills training Functional Skills Training develops key CPAT skill training for lower back, abdominal, balance, and reaction time. Functional skill training enhances all other components because the forces applied by arms and legs must pass through the core of the body.

General adaptation syndrome (GAS) An organism's response to stress. Positive adaptation to stress is called **eustress**, as being the result of correctly timed alternation between stress and regeneration. Too great a stimulus and/or too little regeneration results in a negative adaptation, or **distress**.

Goal setting Goal setting is the foundation on which all of training and practice is designed. It guides what you do, when you will do it, and at what intensity.

Heart rate monitor A device that allows a user to monitor their heart rate while exercising.

High-rise pack A bundle of fire hose, weighing approximately 25 lbs (11.34 kg).

Hydration level The amount of water in the body.

Hydrostatic (underwater) weighing An accurate but expensive and time consuming method for measuring body composition and determining lean body mass (muscle) and body fat. This method involves weighing a person under water, generally in a research facility.

Incumbent firefighters Incumbent firefighters are personnel who are currently in the position.

International Association of Fire Chiefs (IAFC) An international network of over 12,000 fire chiefs and fire department chief officers, established in 1873. Members are the world's experts in fire fighting, emergency medical services, terrorism response, hazardous materials spills, natural disasters, search & rescue, and public safety legislation.

International Association of Fire Fighters (IAFF) Organization representing over 230,000 professional firefighters and emergency personnel.

Kinesiologist A medical specialist who studies muscles and their movements.

Kinesthetic awareness Kinesthetic awareness results when movements have the right "feel" or your performance was "smooth."

Lactate threshold (LT) The point where you start using carbohydrates instead of fats as a primary energy source.

Learning points Key behaviors or techniques that led to success in each CPAT event.

Musculoskeletal Involving both the muscular and the skeletal systems.

National Fire Protection Association (NFPA) A professional organization that advocates and publishes fire and building safety codes and standards and provides educational and fire safety information. The Association publishes the National Fire Codes and the Learn Not to Burn Curriculum.

One repetition maximum (1RM) The maximum amount of weight that can be lifted for one repetition. To calculate 1RM, divide the 10RM weight by 0.75. Divide this by your body weight to determine body weight/bench press ratio.

Overlearning Overlearning entails practicing far beyond the point where you have mastered each task so that it is a reflexive or automatic action.

Overtraining Decreased work capacity resulting from too little rest or too much training.

Periodization A training method that seeks to "peak" an athlete's performance with a competitive event. This is accomplished by changing training intensity and strategically placing maintenance and recovery phases to enhance performance.

Pike pole A long pole with a hook at the end.

Protein A nutrient used to repair tissue damage, build the immune system, and help the body recover from exercise. Proteins are the building blocks for all tissue, enzymes, and hormones that control our movements and metabolism.

Rating of perceived exertion (RPE) A method of assessing exercise intensity.

Reflexive Automatic or without conscious control.

Repetition A successful completion of an exercise movement.

Resting heart rate (RHR) The heart rate indicates your basic fitness level.

Self-control A mental training skill that involves eliminating worry and accepting input and advice from others such as coaches, mentors, and instructors.

Self-talk A mental training skill that involves focusing on positive thoughts to build confidence.

Set A group of repetitions.

Skin fold measurement An indirect means of measuring body composition made by grasping the skin and underlying tissue, shaking it to exclude any muscle, and pinching it between the jaws of a caliper.

Static stretching An exercise that involves going into a stretch position and holding it at your point of limitation for 15–30 seconds. Avoid bouncing as this will tighten muscles.

Strength training Resistance training exercise usually involving weights or some other form of resistance designed toward improving an individual's strength.

Supercompensation An enhanced capability following a positive adaptation to the stress, when the organism is capable of doing more work.

Target heart rate zone The range of heart rates that are optimal for improving fitness.

Uncharged A fire hose line not filled with water.

Valid Well-founded and legally binding.

Visualization A mental training skill that involves mentally experiencing each event as if living it.

INDEX

Note: **b**, **f**, or **t** after a page reference indicates related boxes, figures, or tables.